Darwin's
God

DARWIN'S GOD

EVOLUTION AND THE PROBLEM OF EVIL

Cornelius G. Hunter

Brazos Press
A Division of Baker Book House Co
Grand Rapids, Michigan 49516

© 2001 by Cornelius G. Hunter

Published by Brazos Press
a division of Baker Book House Company
P.O. Box 6287, Grand Rapids, MI 49516-6287

Paperback edition published 2002

Second printing, July 2004

Printed in the United States of America

Library of Congress Cataloging-in-Publication Data

Hunter, Cornelius.
    Darwin's God : evolution and the problem of evil / by Cornelius Hunter.
        p.   cm.
    Includes bibliographical references.
    ISBN 1-58743-011-8 (cloth)
    ISBN 1-58743-053-3 (paper)
    1. Evolution. 2. Evolution (Biology). 3. Evolution—Religious aspects.
    4. Darwin, Charles, 1809–1882. I. Title.
    B818.H855 2001
    213—dc21   00-050761

For current information about all releases from Brazos Press, visit our web site:
http://www.brazospress.com

Contents

Preface

Rarely, it seems, do the stories of history fall into neat categories for our consumption. The American colonies heroically fought their way out of British dominion, but truth be known, most Americans were ambivalent or even against the war for independence. Abraham Lincoln spoke powerfully against slavery in the Lincoln-Douglas debates, but Douglas hated slavery as much as Lincoln. The stories of history require a certain amount of care and attention if we want to really understand them. When the necessary care is missing, the result is an overly simplified and inaccurate rendition. One such story in dire need of some care is science's theory of evolution.

Evolution is more than just a scientific theory, if only because of its tremendous influence in areas outside of science. It is probably the most influential idea ever generated by science. From public policy to the pulpit, and in most things in between, one can find influences from evolution. But despite its influence, evolution is not well understood. Over its first one hundred and forty years, evolution has undergone a great variety of often opposing descriptions. And these descriptions routinely leave out important details, the shades of gray that we need to understand evolution. For some people evolution is nothing more than good science; others say it is bad science. Some say it is a description of reality, a great triumph of Western thought; others say it is philosophically flawed, relying on either tautological or nonfalsifiable ideas. Some say evolution proves materialism, while others say it presupposes materialism. Some say evolution is God's creation tool; others say it excludes God. To be sure, within evolution's broad boundaries one can find instances of many of these things. But at its core, evolution is none of these things.

Evolution is neither atheism in disguise nor merely science at work. There are shades of gray in the story of evolution, and it is these subtleties that reveal the true essence of evolution. The story of evolution is not simple, but it is important, for we need a better understanding of this most influential idea of our time. This book is the story of evolution, including its scientific and its nonscientific aspects. Evolution cannot be fully understood without exploring both aspects, for between them there is considerable overlap and exchange. Scientific reasoning is often beholden to the presuppositions of the day, and this is certainly true for evolution. Hence this is not a story about science or philosophy or religion but rather a story about how all these influences come together in Charles Darwin's theory of evolution.

This book could not have been written without the help of a great many people. By far the most important person has been my wife, Jeanine, who was always ready with assistance and patience. I would also like to thank John Wilson, Daniel Dix, Michael Behe, Kristin Brodowski, Elihu Carranza, Michael Shea, David Fleetham, Rodney Clapp, and Phillip Johnson. I am, of course, solely responsible for any errors in the book.

1

Where Science Meets Religion

I n 1859 Charles Darwin presented his theory of evolution to the world. Although many discoveries have been made since that time, the basic idea behind the theory remains the same today. Darwin proposed that life was the result of an undirected process and that nothing more than the interplay of natural forces was sufficient to produce nature's vast array of species.

Darwin was by no means the first person to advance a naturalistic explanation for life, but he was the first to provide a compelling and sophisticated argument. Similar, though less complete, ideas had been discussed for many decades before 1859. All these ideas were in sharp contrast to the doctrine of divine creation—the belief that an intelligent and all-powerful God had personally created the world and all its living creatures. The doctrine of divine creation had held sway for centuries, but in Darwin's day it faced increasing difficulties. For the most part these difficulties arose from failures to reconcile the Creator with the creation. The Creator was viewed as infinitely wise, powerful, and good. But his creation was increasingly found wanting, with all sorts of defects. Nature was apparently less than perfect. In some regards it was wasteful, and sometimes it seemed downright evil.

Evolution's great success lies in its explanation of the less-than-perfect side of nature. For if the world and even life itself arose from the blind forces of nature, then certainly we should expect a rather imperfect result.

The world of biology is, to be sure, full of beauty and wonder. But there also seem to be anomalies and inefficiencies. Darwin was concerned, for example, that tons of pollen go to waste each year, that some species are ill-adapted for their environments, that ants make slaves of other ants, and that parasites feed off their victims. He tried to make sense of what seemed to be the evil side of nature.[1] "What a book a devil's chaplain might write on the clumsy, wasteful, blundering, low, and horribly cruel works of nature,"[2] he concluded in a letter to a friend.

How could divine creation be reconciled with such evils? It was questions like these that, for Darwin, seemed to confirm that life is formed by blind natural forces. He was motivated toward evolution not by direct evidence in favor of his new theory but by problems with the common notion of divine creation.[3] Creation, it seemed, does not always reflect the goodness of God, so Darwin advocated a naturalistic explanation to describe how creation came about.

New ideas in science often come in response to the failures of older ideas. Albert Einstein's theory of relativity was a remedy for the problems increasingly apparent in Newtonian physics. Likewise, Charles Darwin saw his theory as a solution to problems with the theory of divine creation. The failures of old ideas are important precisely because they are the motivation for new ideas. One cannot fully appreciate the new ideas without first understanding those failures. Indeed, new ideas are *predicated* on the rejection of the older ideas they are replacing—in this sense the rejection of the older idea is a part of the new idea.

This is an important point, because in the case of evolution the older idea is a religious one, not a scientific one. Evolution is predicated on the failure not of a scientific idea but of a religious idea. Is the downfall of this religious idea incidental, or does evolution rely on it for justification? This book will show the latter is true.

There is, to be sure, plenty of evidence supporting evolution, but there is plenty of evidence for all sorts of discarded theories. In fact, one can formulate arguments against evolution, often using the same evidence, that are more persuasive than the supporting arguments. But there is, as we shall see, a line of nonscientific—metaphysical—reasoning that is consistently used to support evolution. It uses scientific observations to argue against the possibility of divine creation. Such *negative theology* is metaphysical because it requires certain premises about the nature of God. A great irony reveals itself here: evolution, the theory that made God unnecessary, is itself supported by arguments containing premises about the nature of God. There is a profound yet subtle religious influence in the theory of evolution. Darwin as well as today's modern evolutionists appeal to these metaphysical arguments.

Evolution cannot be understood if we do not first understand how the metaphysics has influenced the science. Chapters 2 through 4 of this book take up the evidence that is typically presented in support of evolution. These chapters show that the evidence makes evolution compelling only when a specific metaphysical interpretation is attached. Examples show how evolutionary theory implicitly relies on negative theology. Chapter 5 is a historical survey of evolutionists since Darwin who attempted to prove evolution—and we find that evolutionists after Darwin also have consistently relied on nonscientific arguments. Chapter 6 looks at the centuries leading up to Darwin's time and shows how earlier thinkers were influenced by what they believed God ought to do. In particular the problem of evil, both moral and natural, increasingly drove thinkers to distance the Creator from his creation.

Chapter 7 focuses on the nineteenth century and the setting in which Darwin worked. Darwin and his fellow naturalists tried to explain the origin of natural evil; evolution was Darwin's solution. Evolution clearly followed earlier notions that distanced God from creation. Chapter 8 examines the metaphysical thought surrounding evolution. Metaphysical arguments have been used to justify and protect the theory, and in turn evolution has influenced metaphysical thought. But that influence is really just an amplification of the metaphysic that set the stage for evolution in the first place. If one already agrees with that metaphysic, then evolution is compelling; otherwise the theory is a failure. The difference comes down not to scientific arguments but to one's metaphysical presuppositions.

Chapter 9 concludes the book by examining responses to Darwinism, ranging from those who reject evolution to theistic evolutionists and orthodox evolutionists. Understanding these various responses requires understanding how each camp interprets evolution's use of metaphysics. Likewise, meaningful debate between the groups will be possible only when these interpretations are properly acknowledged.

The Modern God

One of Darwin's favorite works of literature was John Milton's *Paradise Lost*.[4] He carried a well-worn copy of the classic on his voyage around the world in the HMS *Beagle. Paradise Lost* was not just a favorite of Darwin; it was immensely popular in Victorian England, to the point of having the status of official Christian doctrine for some.[5]

The main purpose of *Paradise Lost* was to solve the problem of evil. If God is loving and all-powerful, why does he allow evil to exist at all? Milton tried to explain the purposes of God, or as he put it, "justify the ways of God to man." His solution was that God needed to let humans choose between good and evil so he could separate the good from the bad. Although this solution maintained God's purity, it also made him somewhat passive, distanced from the events of history. This epic tale is in many ways a telling signpost of where the modern era was going with its view of God and creation. Creation was on its own, rather than under God's influence and control.

An important similarity between Darwin and Milton should not be missed. The two are sometimes contrasted, since Darwin was rapidly moving toward a naturalistic explanation of the world, whereas Milton saw God as the creative force of the world. But both men were dealing with the problem of evil—Milton with moral evil and Darwin with natural evil—and both found solutions by distancing God from the evil. And most important, the two held similar conceptions of God.

Medieval theologians had attempted to prove the existence of God using logical arguments and the evidence of the created order. In the seventeenth century and after, modern theologians and philosophers internalized and amplified this tradition. Science's new discoveries about nature were fitted into proofs of God. But a God whose existence could be proved was a God who was subject to human reasoning. Increasingly theologians and philosophers felt free to place bounds on what God could do.[6]

This was the dominant view of God amongst modern intellectuals, and it was shared by Darwin and Milton. This link between the two thinkers is more important than their differences. It would be more accurate to view Darwin not as opposing Milton but as extrapolating from Milton. Milton may have justified God, but he did so by distancing God from the moral evils of the world. Darwin, dealing with natural evils, simply distanced God even further. And though Darwin made repeated references to the Creator, he never needed to define his terms, for the modern view of God was widely accepted.

In constructing the arguments for his theory of evolution, Darwin repeatedly argued that God would never have created the world that the nineteenth-century naturalists were uncovering. Shortly after going public with his theory, Darwin wrote to a friend: "There seems to me too much misery in the world. I cannot persuade myself that a beneficent and omnipotent God would have designedly created the [parasitic wasp] with the express intention of their feeding within the living bodies of caterpillars, or that the cat should play with mice."[7]

Darwin had a long list of biological quandaries that did not fit with the view of God that was popular in his day. There was, for example, the problem of hybrids. Why should species cross so easily if they were created separately? And if fauna and flora were specifically created for their environments by a wise Creator, how is it that plants that are introduced into a new region may be successful though they have little in common with the indigenous flora? What seemed to be specialized fauna or flora sometimes flourished in foreign environments. Why were the inhabitants of similar but separate environments, such as cave-dwelling creatures on different continents, often so vastly different? Then there were the ill-adapted species, such as the land animals with webbed feet and the marine creatures with nonwebbed feet. Insects that spent hours underwater differed little from their terrestrial cousins. Why was the water ouzel, a member of the thrush family, so active underwater, and why were woodpeckers found in treeless pampas? And there was that annual "incalculable waste" of pollen.[8]

Nature seemed to lack precision and economy in design and was often "inexplicable on the theory of creation."[9] In addition to this growing list of imperfections and mistakes, Darwin questioned the way the various species were designed. He observed, on the one hand, that different species use "an almost infinite diversity of means" for the same task and that this should not be the case if each species had been independently created by a single Creator.[10] On the other hand, Darwin observed that different species use similar means for different tasks.[11] This too, he argued, does not fit with the theory of divine creation.

What exactly did Darwin expect God's creation to look like? We may never know, but for our purposes the point is that Darwin was significantly motivated by nonscientific premises. He had a specific notion of God in view, and as it had for Milton, that view defined the framework of his thinking. Though biology was young and little was known about how organisms actually worked, Darwin believed he had sufficient evidence to show that God would not have created this world. God's world had to fit into certain specific criteria that humans had devised.

This view was not peculiar to Darwin. Philosophers and scientists had become quite confident in their knowledge of God. This attitude developed over many centuries, and by Darwin's day it was internalized and needed no justification. Today this view continues to be evident in evolutionary literature, from popular presentations of the theory to college-level textbooks.

The Evolution Theodicy

Evolutionists sometimes claim that religious ideas play no role in their theory. Darwin's references to the Creator, they say, were necessary to address the claims of the opposing theory—creation. It is true that Darwin addressed the idea of creation that was influential at the time. But Darwin's criticisms of creation were more than this. They served as de facto arguments for evolution—and as we shall see in chapters 2 through 5, this theme continues in today's evolution literature.

Evolutionists use negative theological arguments that give evolution its force. Creation doesn't seem very divine, so evolution must be true. Evolution is a solution to the age-old problem of evil. The problem of evil states that if God is all-powerful and all-good, then he should not allow evil to exist. For centuries theologians and philosophers have tried to solve this problem. As Milton showed in *Paradise Lost*, moral evil can be explained as the result of human autonomy, but natural evil is more difficult to rationalize. The seventeenth-century philosopher Gottfried Leibniz was interested in the problem of evil. He coined the term *theodicy* for any explanation to the problem. By Darwin's day the list of such explanations was growing. One strategy was to try to show that God was somehow disconnected from creation. Natural evil arose not from God's direction but from an imperfect linkage between Creator and creation.

In addition to natural disasters, fires, and plagues, natural evil can include the vagaries of the biological world. The carnage in nature had always been obvious, but the scientific revolution was revealing it in increasing detail. Also, naturalists were finding the created order to be full of apparent inefficiencies and anomalies. From parasites to extinctions, nature seemed to be less than ideal. This facet of natural evil began to be addressed in the eighteenth century when early theories of evolution appeared. They were crude in detail, but they suggested that nature is best explained as the result not of a divine hand but of some combination of unguided forces.

Darwin's concern with the problem of natural evil is apparent in his notebooks and in his published works. His theodicy had a strong scientific flavor, to the point that most readers lost sight of the embedded metaphysical presuppositions. Where earlier solutions lacked detailed explanations, Darwin provided scientific laws and biological details. But Darwin's general approach followed the earlier attempts. God was constrained to benevolence and was distanced from the evils of creation through the interposition of natural laws. Positing natural selection operating in an unguided fashion on natural biological diversity was Darwin's unique solution. But his overall approach, to distance God from evil, was predictable.

But if God must be distanced from creation's evils, some believed he still must be kept within view to account for morality. Our strong inner sense of right and wrong seems to go beyond personal opinion or preference. For the eighteenth-century philosopher Immanuel Kant, our innate moral sense is sufficient to prove the existence of God. The nineteenth-century geologist Reverend Adam Sedgwick would concur with Kant's conclusion.

Sedgwick was a popular figure at Cambridge in Darwin's day and was known for engaging lectures, which he continued to give until the age of eighty-six. He was at different times president of the Geological Society of London, president of the British Association, and vice-master of Trinity College. Scientists and philosophers have for centuries argued that the created order is proof of an almighty Creator. Sedgwick agreed but, following Kant, also found such proof within our own morality: "In the material world we see in all things the proofs of intelligence and power; so also, that in the immaterial world we find proofs, not less strong, that man is under the moral government of an all-powerful, benevolent, and holy God."[12]

Sedgwick consistently focused on morality and its link to the natural sciences. He believed that God governed by general, fixed laws in both the moral and physical worlds. This moral imperative, for Sedgwick, meant that we must keep God in view. He believed that exploring the created order is a privilege for naturalists, which they should not abuse by denying the divine hand behind creation. Sedgwick denounced pre-Darwinian theories of evolution, calling them nothing better than a "phrensied dream."[13]

In one of his summer field expeditions to Wales, Sedgwick took on the young Charles Darwin. It was a good experience for Darwin, and the two men remained friends, but as Darwin matured in his studies of nature he increasingly viewed nature as anomalous, inefficient, and downright brutal. How could an all-good God create such a gritty reality?

The problem was aggravated by the rather two-dimensional God the Victorians had in view. It was a tradition that had been building for centuries, and by Darwin's day the popular conception of God was a very pleasant one. Positive divine attributes such as wisdom and benevolence were emphasized to the point that God's wrath and use of evil were rarely considered.

Few people promoted this doctrine of God more avidly than the orthodox Sedgwick. Sedgwick often spoke of God's power, wisdom, and goodness. His main point of application was how these positive attributes are manifest in creation. The student of nature, according to Sedgwick, should find the natural world full of beauty, harmony, symmetry, and order. Biology was full of beautiful form and perfect mechanism "exactly fitted to the vital functions of the being."[14] And it was all driven by God's wonderful laws: "What are the laws of nature but the manifestations of his wisdom? What are the material actions but manifestations of his power? Indications of his wisdom and his power co-exist with every portion of the universe. They are seen in the great luminaries of heaven—they are seen in the dead matter whereon we trample."[15]

15

According to Sedgwick, nature was never anomalous or fortuitous. Sedgwick's idealism was as apparent in what he did write as in what he did not write. When he quoted Scripture he consistently avoided the passages that link God and evil. Sedgwick quoted the passage in Job where God reveals his power but not the passage where God reveals that the ostrich treats her young harshly because he has deprived her of wisdom. He quoted the Romans passage where Paul revels in God's eternal power and how it is reflected in creation but not the passage where creation groans because God has subjected it to futility. He quoted the psalmist's proclamation that creation declares God's glory but not Isaiah's prophecy of how God creates calamity.[16]

Sedgwick and his generation had rather idyllic expectations for the natural world. What was a young naturalist like Darwin to think when he found parasites slowly torturing their hosts? Nature was turning out to be less pretty than Sedgwick had predicted, and Darwin searched for an explanation. His solution was to distance God from creation by interposing a natural law—his law of natural selection.

Darwin's theory of evolution was very much a solution to the problem of natural evil—a theodicy. The problem had confounded thinkers for centuries. They needed to distance God to clear him of any evil doings. Darwin solved the problem by coming up with a natural law that he argued could account for evil. Natural selection, operating blindly on a pool of biological diversity, according to Darwin, could produce nature's carnage and waste.

Darwin's solution distanced God from creation to the point that God was unnecessary. One could still believe in God, but not in God's providence. Separating God from creation and its evils meant that God could have no direct influence or control over the world. God may have created the world, but ever since that point it has run according to impersonal natural laws that may now and then produce natural evil.

Darwin may have solved the problem of how nature's evil could arise without God, but what about Sedgwick's morality? Though he respected Darwin's effort, Sedgwick criticized his theory of evolution scathingly. Predictably enough, the main complaint was that by distancing God, Darwin was disregarding the moral imperative that was so obvious to Sedgwick. After reading Darwin's book, *The Origin of Species*, Sedgwick wrote to Darwin:

> There is a moral or metaphysical part of nature, as well as a physical. A man who denies this is deep in the mire of folly. 'Tis the crown and glory of organic science that it does through final cause, link material and moral; and yet does not allow us to mingle them in our first conception of laws, and our classification of such laws, whether we consider one side

of nature or the other. You have ignored this link; and, if I do not mistake your meaning, you have done your best in one or two pregnant cases to break it.[17]

After the sixteenth century, modernism had tended to view God as removed from creation, but Darwin was now increasing this separation to the point that the link between creation and God was severed. Sedgwick disagreed with Darwin's removal of God, not with Darwin's conception of God. Sedgwick complained that Darwin had removed the moral imperative, but he failed to see that it was modernism's doctrine of God—a highly constrained view of God that he and Darwin shared— that drove Darwin's reasonings.

Sedgwick reiterated the popular idealism of the day: "I can prove," he wrote, that God "acts for the good of His creatures."[18] This sort of thinking simply fueled Darwin's theory. For if God acts for the good of his creatures and those creatures are sometimes found to be in dire straits, then God's acts must have been hindered along the way. If God did not act directly upon creation but instead installed a law such as natural selection as its governor, then creation's scars could be explained.

The difference between Sedgwick and Darwin, then, lay not in their conception of God but in the metaphysical problem that colored their study of nature. Sedgwick was concerned with morality, and Darwin was concerned with evil. Darwin, it seemed to Sedgwick, had removed the moral authority—but did this mean Darwin was blind to nature's moral implications, as Sedgwick charged? Not at all; in fact, it was nature's cruel and wasteful aspects that especially concerned him. Darwin was having difficulty reconciling the modern God with what he saw in nature. If a benevolent God manifested himself in an idyllic world, then why was Darwin seeing so much imperfection? Darwin's gritty and chaotic world—the real world seen up close by naturalists—implied no such Creator. Creation was irrational, and therefore there was no such benevolent Creator, or at least not one who attended to details. Darwin summarized this metaphysical argument that underlies evolution in his autobiography:

> Suffering is quite compatible with the belief in Natural Selection, which is not perfect in its action, but tends only to render each species as successful as possible in the battle for life with other species, in wonderfully complex and changing circumstances.

> That there is much suffering in the world no one disputes. Some have attempted to explain it in reference to human beings, imagining that it serves their moral improvement. But the number of people in the world is nothing compared with the numbers of all other sentient beings, and these

often suffer greatly without any moral improvement. A being so powerful and so full of knowledge as a God who could create the universe is to our finite minds omnipotent and omniscient. It revolts our understanding to suppose that his benevolence is not unbounded, for what advantage can there be in the sufferings of millions of lower animals throughout almost endless time? This very old argument from the existence of suffering against the existence of an intelligent First Cause seems to me a strong one; and the abundant presence of suffering agrees well with the view that all organic beings have been developed through variation and natural selection.[19]

Darwin's reconciliation resolved the metaphysical dilemma that bothered him but not Sedgwick—the problem of evil. But now, with one metaphysical dilemma gone, another stepped in to take its place—the one that bothered Sedgwick: the problem of morality. What is the source of our moral law?

Sedgwick and Darwin were opposed over a deep theological question that cannot be resolved with a scientific theory. The question had already been debated many times. The terminology changes over time, but the core issue remains the same. *The existence of evil seems to contradict God, but the existence of our deep moral sense seems to confirm God.*

Evolution takes a position on this ancient question, but in doing so it becomes much more than a scientific theory. It is a description of reality based on a metaphysical presupposition, and as such it makes truth claims as no scientific theory can. Whether one accepts or rejects evolution depends in large measure on whether one accepts or rejects its presuppositions. But regardless, we cannot understand evolution without understanding its presuppositions and how it uses them.

2

Comparative Anatomy

he classification of living organisms is an age-old problem.
Meno told Socrates that bees, "insofar as they are bees, do not
differ at all, the one from the other," but that was not much of
a start. Before Darwin, the traditional method of classification
was hierarchical. All known species were divided into broad
categories, such as plants versus animals. These categories were
then subdivided into increasingly smaller and more specific categories.
For example, within the animal category species with a backbone-like
structure were called the chordates; chordates with a spinal column were
called vertebrates; vertebrates that breastfed their young were called
mammals; mammals that were flesh-eating were called carnivores;
carnivores that resembled the dog (wolves, foxes, etc.) were called
canines; and finally canines were divided into genera, and genera were
divided into species.

This method of classification was championed in the eighteenth century by Linnaeus. It reflected the belief that was becoming increasingly prevalent in the modern era, that God operates via laws and decrees rather than particulars. The species found in nature were often viewed as harmonious parts of God's perfect plan that could be organized into a hierarchy of forms. But as more and more species were collected from around the world, the plan grew increasingly complex. Some species simply did not fit neatly into the hierarchy.

With the acceptance of Darwin's theory of evolution it was natural to classify organisms according to their evolutionary relationships. The term *phylogeny* refers to these hypothetical evolutionary relationships. *Phylogeny* is one of the many terms coined by nineteenth-century Darwin disciple Ernst Haeckel (1834–1919). A phylogenetic tree, popularly known as an evolutionary tree, illustrates the supposed branching relationships among species—the evolution of younger species from older species.

But before any phylogenetic tree can be constructed, the species must be compared for differences and similarities. Naturalists have traditionally noticed that such comparisons seem to fall into two different categories. In the first category, birds and bees, for example, both have wings for the same purpose—flight—but common purpose seems to be where the similarity ends. Birds and bees and their wings are otherwise very different. Taxonomists refer to these two wing structures as *analogous*.

With his theory of evolution, Darwin proposed an objective definition for analogous similarities. Birds and bees must be on very different limbs of the evolutionary tree because they are so different. Their most recent common ancestor—somewhere low on the trunk—is far down in the early phases of evolutionary history, long before there were wings. Therefore birds and bees evolved their wings independently of each other. According to Darwin, *analogous structures are those that evolved independently.* Many similarities are considered analogous.

On the other hand, the lizard, the bat, and the human, to name a few, all have five digits (four fingers and a thumb for humans) at the end of the forelimb structure. This common configuration, known as the pentadactyl structure, is found in the fossil record as well and is thought to be common to an entire group, including the original ancestor. Patterns such as this that persist throughout a whole group are referred to as *homologous*.

Before Darwin, homologies were interpreted as a sort of divine template revealing the Creator's unity of design. But Darwin gave homologies a completely new interpretation. *Homologous structures are those that did not evolve independently.* According to Darwin, instead of representing a divine plan, homologies are leftovers of descent with modification. The structure was good enough to do the job, so an entirely new design was not selected. It was easier for evolution to modify an existing plan rather than redesign from scratch.

Within a group of species, similarities that appear to be homologous help evolutionists to reconstruct the group's phylogeny since such similarities suggest evolutionary relationships.[1] Homologies are manifest in different ways. There are useful organs, less useful (vestigial) organs, embryonic homologies, and biochemical homologies. Darwin was aware of all but the biochemical homologies, and the evidence from homology was a critical part of his argument. In fact, Darwin felt that this evidence alone was sufficient to advance his theory,[2] and it remains one of the major pillars of evidence for evolution.[3] This chapter reviews this evidence for evolution, then explores problems with the evidence, and finally shows that the strength of the evidence lies in its metaphysical interpretation.

The Evidence

The pentadactyl structure is found throughout the tetrapods— amphibians, reptiles, birds, and mammals with two sets of limbs. The activities of this group of fauna include flying, grasping, climbing, and crawling. Such diverse activities, evolutionists reason, should require diverse kinds of limbs. There seems to be no clear functional or environmental reason that all should need a five-digit limb. Why not three digits for some, eight for others, thirteen for some others, and so forth? Yet they all are endowed with five digits. Bird and bat wings are based on the five-digit limb, albeit in different ways. Seal flippers are pentadactyl limbs, and the boneless hind fin of the whale conceals vestiges of the characteristic five-digit pattern.[4]

Vestigial Organs

Evolutionists believe that certain homologous structures have, over the course of evolution, lost their original purpose. The list of such organs in humans might include wisdom teeth, coccyx (tail vertebrae), ear-wiggling muscles, and appendix.

The modern whale provides an example of what evolutionists believe to be a vestigial structure that may have taken on a new function. Whales lack external hind limbs, but they do have a small set of bones that appear to be homologous with the pelvis of other tetrapods. These bones may serve to support the whale's reproductive organs, but the bones are no longer used for hind limb control. As the whale evolved from ancient tetrapods, evolutionists believe, the pelvis continued to decrease, leaving only the vestiges we see in the modern whale. This structure may not be useless, but it no longer serves its original purpose.[5]

Even more persuasive for evolutionists are vestigial organs that, beyond an apparent lack of function, appear to be inefficient. For example, the recurrent laryngeal nerve connects the brain to the larynx via a tube near the heart. In fish the route taken by the nerve is direct, but in the giraffe the route is circuitous, proceeding down and up the neck. This strange arrangement requires the nerve to be ten or more feet longer than if the shortest path were taken. Of course, the nerve functions just fine in the giraffe, and so the blind process of evolution let it be. Such an inefficiency is easy to explain if giraffes have evolved in small stages from a fishlike ancestor.[6]

At the molecular level, evolutionists believe they have identified vestigial structures in the form of *pseudogenes*—DNA sequences that resemble genes but appear to be nonfunctional. Evolutionists believe pseudogenes are the remnants of ancient genes, no longer in use but carried along as excess baggage.

Ontogeny

Evolutionists also find homologies in the developing stages of life where organisms construct themselves. This self-construction process—referred to as *ontogeny*—can be quite elaborate, taking the newly formed individual through a series of transformations that are very different from each other and from the ultimate adult form.

Of course, no matter how complex the ontogeny, all life starts out as a single cell—the *zygote*. The zygote marks the beginning of the embryonic phase of development, when the animal does not actively feed. The zygote divides into daughter cells, which form a hollow structure called the *blastula*. Next, as the cells continue to divide, one side of the blastula pushes inward to form a cupped structure called the *gastrula*. Up to this point the daughter cells are all identical, distinguished only by their location in the structure. Swap any two cells and the development of the adult organism—the morphogenesis—is unaffected. But after the gastrula stage, cells begin to *differentiate*. They take on different properties that determine their ultimate fate in the organism. Swap two cells at this point and an organ may end up in the wrong place.

As we trace back through earlier phases of development of an animal, the embryo is less differentiated. At the early stages, we may speak in general terms, for most animals appear similar at this point. It is indeed remarkable how similar the early stages of ontogeny are for organisms that are quite dissimilar in the adult stage. According to the nineteenth-century embryologist Karl Ernst von Baer, "the embryos of mammalia, of birds, lizards, and snakes . . . are in their earliest states exceedingly like one another, both as a whole and in the mode of development of their parts; so much so, in fact, that we can often distinguish the embryos only by their size."[7] Darwin referred to this as the "law of embryonic resemblance."[8] He argued it was obvious that such similarities did not arise from any functional requirement—for we have no more reason to believe this than "to believe that the similar bones in the hand of a man, wing of a bat, and fin of a porpoise, are related to similar conditions of life."[9]

That these early embryonic forms, so different from the subsequent adult forms, are not necessary is evident in cases of *gradual metamorphosis*—species whose embryos resemble miniature adults. If some species could have such embryos, Darwin reasoned, then "there is no reason why, for instance, the wing of a bat, or the fin of a porpoise, should not have been sketched out with all their parts in proper proportion, as soon as any part became visible."[10]

To summarize, very different animals have similar embryos that apparently are not designed for their respective unique requirements. Darwin concluded that the resemblances were homologous. Furthermore, he suggested, they reveal to us the structure of the evolutionary ancestors:

> As the embryo often shows us more or less plainly the structure of the less modified and ancient progenitor of the group, we can see why ancient and extinct forms so often resemble in their adult state the embryos of existing species of the same class. Agassiz believes this to be a universal law of nature; and we may hope hereafter to see the law proved true.[11]

What a boon this would be for evolutionists. If true, the law could yield a plethora of data, for the evolutionary ancestors of a species would be manifest in its earliest stages of life. Where the ancient fossil record had its gaps, perhaps modern embryos could fill in the picture.

This notion set the stage for Haeckel's famous dictum "ontogeny recapitulates phylogeny," otherwise known as the biogenetic law.[12] In its strong form, it states that the early development of an individual is a brief and rapid review of its evolutionary history. For example, the zygote, blastula, and gastrula of an advanced organism could reflect the primitive bacteria and its colonies from which the organism ultimately was derived. In humans, the embryo eventually reaches a stage that includes structures resembling the gill structures (aortic arches) of fishes, and much later there is a furlike coat that temporarily covers the fetus.

23

There is also, for example, the case of the mammalian kidney, which goes through three stages of embryological development. The first stage, the pronephros, resembles the kidney of primitive fishes. The second stage, the mesonephros, resembles the kidney of more advanced fishes and amphibians. The third stage, the metanephros, is similar to the adult kidney found in all terrestrial reptiles, birds, and mammals. These three stages could be homologous with the adult kidneys of the evolutionary ancestors, in the correct evolutionary order.

But the strong form of Haeckel's law has long since been disproved. Even Haeckel found many exceptions.[13] Still, if embryos do not repeat ancestral *adult* stages, they may yet repeat ancestral *developmental* stages. The "gills" of the human embryo are viewed as homologous to those of the fish embryo, which persist to become the gill slits of the adult fish.

Evolutionists do not regard phylogeny as in some way causing ontogeny. Instead, just as evolutionary histories are to a certain extent evident in the adult form, they are also evident in the immature stages. Thus in the weak form of Haeckel's law ontogeny does not recapitulate phylogeny, but it may reveal it by recapitulating an ancestral ontogeny.[14] The conclusions of evolutionist Tim Berra illustrate today's view:

> There are many features of embryonic development common to related animals, and the closer the relationship, the more similar the development. The early embryos of all vertebrate classes (fishes, amphibians, reptiles, birds, and mammals) resemble one another markedly. The embryos of vertebrates that do not respire by means of gills (reptiles, birds, and mammals) nevertheless pass through a gill-slit stage complete with aortic arches and a two-chambered heart, like those of fish. The passage of a fishlike stage by the embryos of the higher vertebrates . . . is readily accounted for as an evolutionary relic.[15]

The Universal Genetic Code

The rapid growth of molecular biology in the twentieth century added a new category of homologies—those at the biochemical level. At this level evolutionists find homologies with the widest distribution possible: they are in all life. In fact, it has produced what may be the ultimate homology, the DNA (deoxyribonucleic acid) macromolecule and its universal genetic code. The cell machinery uses the genetic code to interpret the information stored in the DNA molecule when creating proteins.

Virtually all living organisms, from primitive bacteria to plants to animals, make use of the same code. For evolutionists, the universality of the genetic code is important evidence that all life shares a single origin.[16] There is no apparent reason for the particular code that we find.

We might envision certain constraints from chemistry that would make one code preferable over the alternatives, but no such constraint has been found. Instead the code appears to be arbitrary in the same sense that human language is arbitrary (we all agree what the words *tree* and *shrub* mean, but if we all agreed to reverse the meanings, the words would work just as well). Evolutionists therefore view the code as the result of a historical event, what DNA codiscoverer and Nobel laureate Francis Crick called a "frozen accident." The code, they say, originally evolved as the result of blind forces, but once established, it was strongly maintained.

Molecular Comparisons

Evolutionists also use molecular homologies to hypothesize evolutionary relationships between species. In the 1960s molecular biologists learned how to analyze protein molecules and determine the sequence of amino acids that makes up a protein. It was then discovered that a given protein molecule varies somewhat from species to species. For example, hemoglobin, a blood protein, has similar function, overall size, and structure in different species, but its amino acid sequence varies between species. Emile Zuckerkandl and Linus Pauling hypothesized that such sequence differences resulted from a relatively constant rate of evolutionary change occurring over the history of life and could be used to estimate past speciation events—a notion that became known as the *molecular clock*.[17]

For example, if two species have hemoglobin proteins that are almost identical, evolutionists infer that the two species have a recent common ancestor on the evolutionary tree. Only recently did the two species diverge, given that their hemoglobin proteins are so similar. On the other hand, if the two hemoglobin proteins have many differences, evolutionists believe that the two species have been evolving independently for a longer time. The most recent common ancestor of the two species would be lower on the evolutionary tree. The National Academy of Sciences claims that the molecular clock "determines evolutionary relationships among organisms, and it indicates the time in the past when species started to diverge from one another."[18]

In addition to the molecular clock, phylogenies based on molecular comparisons have been claimed to support evolution. One such claim that has been cited in evolution textbooks comes from David Penny and coworkers. Penny used five proteins (cytochrome c, hemoglobin:A, hemoglobin:B, fibrinopeptide:A, and fibrinopeptide:B) to infer evolutionary relationships among eleven different species (rhesus monkey, sheep, horse, kangaroo, mouse, rabbit, dog, pig, human, cow, and ape).[19]

25

There are millions of different ways that eleven species could be arranged in an evolutionary tree. Penny used protein comparisons among the species to judge which of the arrangements would be more likely if they indeed were related via evolution. Penny repeated this process five times, once for each protein, and he obtained similar results. That is, the most likely phylogenies suggested by the five different proteins were all similar. It seemed that five independent measures all produced a similar result. If evolution is not true, Penny argued, then isn't it a strange coincidence that these different proteins all suggest the same pattern? For evolutionists this seems like a resounding confirmation of evolution.[20]

High Confidence in Molecular Evidence

Evolutionists believe that the fruits of molecular biology, unknown to Darwin, have resoundingly confirmed his theory. In fact, it is difficult to overestimate evolutionists' confidence in biochemical homologies. According to the National Academy of Sciences, the "evidence for evolution from molecular biology is overwhelming."[21] For philosopher Michael Ruse "molecular biology has opened up dramatic new veins of support" for evolution, and the theory is now beyond reasonable doubt. "The essential macromolecules of life speak no less eloquently about the past than does any other level of the biological world."[22]

Christian de Duve starts out his ambitious history of life with a triumphant declaration:

> Life is one. This fact, implicitly recognized by the use of a single word to encompass objects as different as trees, mushrooms, fish, and humans, has now been established beyond doubt. Each advance in the resolving power of our tools, from the hesitant beginnings of microscopy little more than three centuries ago to the incisive techniques of molecular biology, has further strengthened the view that all extant living organisms are constructed of the same materials, function according to the same principles, and, indeed, are actually related. All are descendants of a single ancestral form of life. This fact is now established thanks to the comparative sequencing of proteins and nucleic acids.[23]

For Kenneth R. Miller, the 100-percent match in certain pseudogenes is *proof* that humans and gorillas evolved from a common ancestor.[24] For Niles Eldredge, molecular biology provided a rigorous test that evolution successfully passed: "The basic notion that life has evolved passes its severest test with flying colors: the underlying chemical uniformity of life, and the myriad patterns of special similarities shared by smaller groups of more closely related organisms, all point to a grand pattern of 'descent with modification.'"[25]

But despite all the accolades, the argument from homology, including the molecular homologies, is problematic. We have thus far reviewed how the concept of homology is used to support evolution; we now turn to the opposing perspective. As we shall see, though the argument is important to evolutionists, it also raises serious doubts about the theory.

Problems with the Evidence

The first problem with the evolutionist's homology argument is its generality. One of the criteria for judging scientific theories is specificity. In order to be useful and testable, a theory should make predictions of what will and will not occur, and the more specific the predictions the better. Theories that make general predictions that can accommodate just about any result not only are less useful but are protected from falsification.

The homology argument is quite general, for it says that any pattern found in nature was produced by evolution. Nature has millions of different species with incredible diversity. There are species that live high in the mountains, at the bottom of the sea, in arctic conditions, in the rain forest, in the desert, and so forth. Evolution, we are told, produced organisms for all of the different environments and habitats that nature has to offer. There is a tremendous diversity of designs and behaviors that one finds in nature's species. Even in a given habitat there is a great variety of species.

Homologies Cannot Be Unambiguously Identified

Evolution is supposed to have created all this diversity. It seems to be capable of designing and implementing every conceivable biological design. Yet on the other hand, when a pattern is found—a similarity between species—this is supposed to be an example of how stingy evolution can be. Evolution, we are told, favors practicality over optimality. Instead of designing the perfect species, it uses spare parts that are available from ancestral species. On the one hand, evolution seems to have tremendous creative powers, bringing forth the millions of species with all their diversity; yet on the other hand, it is pragmatic. It exerts its creative powers only to the extent that is necessary, often settling for less than optimum designs in the name of expediency.

Although evolution is claimed to be capable of producing whatever we find in nature, there is no objective and unambiguous method for identifying the purported homologies. The pentadactyl pattern discussed above seemed obvious, but nature often fails to present us with such clear-cut cases. In fact, evolutionists caution themselves that homologous structures need not look alike. Mere resemblance does not qualify as a criterion for the very reason that any two homologous structures have gone down their own respective evolutionary paths and in the process may have taken on different appearances.

So how are evolutionists to recognize homologies? Unfortunately there is no set procedure to follow. Instead homologies are analyzed on a case-by-case basis. There are tests to apply, but success is never guaranteed. So evolutionists use a process of elimination to try to distinguish homologies from analogies. First, they compare the fundamental structures to ensure anatomical similarity. Second, they compare the relationship to surrounding characters—the *positional criterion*. Bones must be connected in the same way, and homologies must have the same relationship to surrounding characters. And third, they compare the development patterns. Homologous structures must not just be similar in the adult form but also should have a common embryological development pattern.[26]

This process of elimination still leaves a mixed bag, because while it may detect some analogies that at first appeared as homologies, it does not necessarily detect all of them. In fact, the process described above is a bit idealistic. For example, similarity in early embryology may sometimes be a false requirement, because evolution could modify the development patterns of homologous organs just as it can modify the mature forms.

Mark Ridley gives the following hypothetical example in his textbook on evolution:

> Maybe, from that initial list of 100 characters [traits], morphologic research will cut it down to a list of 30 reliable characters (i.e., can be reliably counted as derived homologies in so far as morphology is concerned), and of them perhaps 20 point to one phylogeny, eight to another, and two to a third. Then we have to fall back on the principle of parsimony [the assumption that the true phylogeny requires the smallest number of evolutionary changes]. A phylogeny treating the 20 characters as homologies and the other 10 as analogies will require fewer evolutionary events than one treating the eight as homologies and the other 22 as analogies. *The analysis by parsimony itself is a criterion of homology,* for it tells us which of the 30 characters were most likely to be inherited from common ancestors and which to be convergent.[27]

We see here that in order to identify homologies we must use the principle of parsimony; otherwise our identification will be subjective. But aside from invoking evolutionary theory, we have no justification for the principle of parsimony. As Ridley points out, that principle is itself a criterion of homology.

At this point we must stand back and ask how homologies can serve as powerful evidence for evolution yet rely on evolution for their very identification. In fact homologies do not provide the resounding confirmation for evolution in the way that, for example, the successful prediction of a rocket trajectory provides confirmation for gravitational theory. Rather, evolution merely accommodates homologies.

Thus in his discussion of homologies Darwin sounds slightly less than triumphant when he concludes: "On the theory of natural selection, we can, *to a certain extent*, answer these questions."[28]

Amazing Analogies

The problem is that without an objective method for parsing out homologies from analogies, we never know when we have wheat or we have chaff. And often nature fails to help us. Instead of making things obvious, it leaves us with difficult decisions. Sir Gavin de Beer, British embryologist and past director of the British Museum of Natural History, has well documented the problems of tracing homologous structures to their initial, embryonic development. For example, two closely related species of frog, *Rana fusca* and *Rana esculents*, have eye lenses that are similar but are formed differently. Did these two cousin species evolve their eye lenses independently? This was not an isolated example, and de Beer concluded, against the third criterion in the above process of elimination, that candidate homologous structures *can* develop differently without forfeiting their homology.

In other cases we are presented with too many candidate homologies. If species A shares similarities with species B and C, but B and C are not closely related, then A can be closely related to B or C, but not both. If the similarities in species A and B are homologous, for example, then the similarities in species A and C would have to be analogous—a case of what evolutionists call *convergent* evolution. Sometimes these cases of convergent evolution produce remarkably similar species, and we are left to wonder how the unguided process of evolution could ever hit upon the same design twice.

Consider the streamlined torpedo shapes, tall dorsal fins, and broad tails found in sharks, swordfishes, and dolphins. None of these are closely related, because they belong in disparate groups (sharks with the cartilaginous fishes, swordfishes with the bony fishes, dolphins with the mammals). Therefore evolutionists believe they are only distantly related, and so their similarities must be analogous, not homologous. Similarly, there is the case of the desert flora, as Berra explains:

> In the southern African deserts are succulent plants of the spurge family, Euphorbiaceae, that look for all the world like the cacti of the Western Hemisphere, but have wholly different ancestors. In both cases, there are tall spire-like species, barrel-shaped species, shrubby species, ground-hugging species, and still others, each adapted to a particular desert ecological niche, but their floral parts and other basic features show that the American cacti (family Cactaceae) and the South African euphorbs are unrelated.[29]

Another well-known example is the marsupial-placental convergence in mammals. In marsupials the young are born soon after conception and continue their early development in a pouch on the mother's belly for weeks or months. In the placentals the embryo is attached to the uterus wall and grows considerably before birth. All mammals are believed to have evolved from a small, four-footed animal something like today's mouse or shrew. The placentals and marsupials are supposed to have formed an early division in the ancestral lineages derived from this humble creature.

The marsupial and placental characters in mammals are homologies for their respective lineages. And, of course, these lineages are impressive radiations. They include hundreds of distinctive types, including the bat, the whale, the wolf, and the rodent. There are creatures that fly, glide, climb, swim, dig, and graze. Some eat plants, some are meat-eaters, and some do both. But among this tremendous diversity there are uncanny similarities in lineages that otherwise divide between the marsupial and placental lines.

For example, there are striking similarities between the marsupial and placental saber-toothed carnivores—*Thylacosmilus* of South America and *Smilodon* of North America. Both sport the same distinctive stabbing upper canines and a protecting flange of bone on the lower jaw. The marsupial known as the "flying phalanger" and its North American counterpart the flying squirrel have distinctive coats that extend from the wrist to the ankle, giving them the appearance of a living hang glider. This along with bushy tails give them their gliding abilities. Then there is the marsupial mole of Australia, whose body plan and behavior are like those of moles of the Northern Hemisphere. Both have enlarged forelimbs and reduced eyes for their subterranean environment. The marsupial and placental wolves have very similar skull shape and body form. And there are more. One can find marsupial counterparts to the placental rats, anteaters, cats, and mice.

All these "cousin" species have their differences, especially in child-bearing habits, but the repeated duplication of distinctive characters is remarkable. Nonetheless, evolution must consider these to be cases of convergent evolution. For example, though marsupial and placental wolves have their similarities, an evolutionary classification of the marsupial wolf makes it more closely related to the kangaroo. We must not be misled, evolutionists warn us, by the striking similarities between the marsupials and the placentals.

There are also amazing analogies at the molecular level. The text-book example of this is subtilisin, a bacterial enzyme that chops up other protein molecules. The function of subtilisin is intricate. It is assembled so that three of its amino acids, which are initially far apart, come together in a precise geometry to enable fast and efficient enzymatic reactions. The reactions are characterized by several key functions that act in concert. There are other enzymes that perform this same set of functions, and so they and subtilisin can be categorized together. But aside from its enzymatic functions, the subtilisin molecule is sufficiently different from the others that it is generally thought to have evolved independently. On the other hand, it could be argued that subtilisin is a case of divergent evolution—that for some reason it evolved to be significantly different from its cousin enzymes but they all share the same ancestor molecule.[30]

Of course, evolution can explain all of this, but the explanations go no further than sweeping generalities about how evolution "in due course selects the most efficient design for the animal's or plant's lifestyle in the particular set of the environmental circumstances."[31] In fact, evolutionists seem to be quite content with this explanation. As Berra concludes: "Such close similarities in very unrelated groups are easily explained as a result of convergent evolution."[32] Perhaps too easily. Though evolution is a blind process that produces a broad menagerie of species and designs, it is also supposed to produce striking similarities.

The Problem of Measuring Fitness

But if there is some uncertainty about those homologies that serve useful functions, what about the vestigial organs—those homologous characters that seem to serve little or no purpose, or perhaps are even inefficient? While the argument from vestigial organs appears persuasive, it too suffers from the lack of an objective measure. The problem is that in order to identify an organ as vestigial, we need to measure its adaptive value—its contribution to the production of offspring. At the

31

core of evolutionary theory is Darwin's law stating that in most instances it is the fittest that reproduce. But due to the complexities of nature and its life forms, we usually cannot measure fitness aside from counting offspring. Those organisms that leave more offspring are usually more fit, but we are not sure precisely why.

Thus it is difficult to show that a particular organ lacks value. Whether we are talking about an organ that is thought to contribute little to overall fitness or one thought to be inefficient, our failure to find positive value does not imply that it is nonexistent. One cannot conclude something does not exist unless one has looked in all possible places at all possible times. In fact, the claim that an organ is vestigial can only be rejected. When we find that the organ makes a positive contribution to fitness, then we disprove the vestigial claim, but it is practically impossible to prove the claim by failing to find such a contribution. It is not surprising therefore that the history of vestigial organs involves shrinking lists.

In 1895 Ernst Wiedersheim published a list of eighty-six organs in the human body that he supposed to be vestigial. The vast majority of items on Wiedersheim's list are now known to be functioning organs. The pineal gland, for example, is now known to be part of the endocrine system that sends chemical messages (hormones) in the blood and interacts with the nervous system. Wiedersheim also claimed the coccyx, a short collection of vertebrae at the end of the spine, was vestigial. But the coccyx is the attachment point for several important muscles and ligaments. And Wiedersheim claimed the thyroid and thymus glands and the appendix were vestigial, but important functions for all three have since been discovered.

The thyroid gland, consisting of two lobes on either side of the windpipe, produces thyroxine, which regulates cellular metabolism. It is important in cold temperatures and in child growth. The thyroid gland also produces calcitonin, which helps regulate blood calcium levels. Its malfunction and enlargement—the disease known as goiter—is visible as a swelling of the front of the neck. Both the thymus gland and the appendix contribute to the body's immune system.

In 1981 zoologist S. R. Scadding analyzed Wiedersheim's claims and had difficulty finding a single item that was not functional, although some are so only in a minor way.[33] He concluded that the so-called vestigial organs provide no evidence for evolutionary theory.

The Vestigial Structure Argument Is Subjective

But even the finding of a positive contribution does not necessarily disqualify a structure from vestigial status, for evolutionists point out that vestigial structures need not be functionless or even inefficient. Evolutionists Edward Dodson and Peter Dodson explain:

> When structures undergo a reduction in size together with a loss of their typical function, that is, when they become vestigial, they are commonly considered to be degenerate and functionless. Simpson has pointed out that this need not be true at all: the loss of the original function may be accompanied by specialization for a new function. Thus the wing of penguins has become reduced to a point that will not permit flight, but at the same time it has become a highly efficient paddle for swimming. The wings of rheas, ostriches, and other running birds are also much reduced, and have been described as "at the most still used for display of the decorative wing feathers." But Simpson has observed that the rheas, when running, spread the wings and use them as balancers, especially when turning rapidly. It seems quite probable that this is true of the running birds generally.[34]

What then defines a vestigial structure? If a penguin's wing is highly efficient for swimming, then why should we think it is vestigial, aside from simply presupposing it was formed by evolution? The idea that vestigial structures can in fact be perfectly useful makes the argument subjective. A character trait that is fully functional for one observer may be only partially functional for another observer, and may be considered inefficient by yet another observer. And so we are again left with evidence for evolution that is subjective.

Berra tells us that the small bones found in rear quarters of whales and snakes are "surely of no value" and that this supports "the evolutionary explanation that whales evolved from terrestrial mammals, and snakes from lizards."[35] Ridley assures us that the recurrent laryngeal nerve "is surely inefficient."[36] But there is no scientific evidence to back up these claims.

The very use of the term *vestigial* begs the question, for vestigial structures serve as evidence for evolution only if they are indeed vestigial. But we cannot know they are vestigial without first presupposing evolution, because we cannot directly measure their contribution to the organism's fitness. Therefore when evolutionists identify a structure as vestigial, it seems that it is the theory of evolution that is justifying the claim, rather than the claim justifying the theory of evolution.

The Embryology Argument Is Subjective

If claims regarding vestigial organs are subjective, can evolutionists at least fall back on one of Darwin's favorites, the argument from embryology? We saw above that the strong form of Haeckel's biogenetic law was replaced by its weak form. Recall that according to the strong form an organism's developmental stages manifest the *adult structures* of its ancestors, whereas the weak form states that an organism's development *may* reflect the *developmental stages* of its ancestors. The reader may have noted that the move to the weak form, while necessary, weakened the evidence. No longer could evolutionists apply a universal law to the observations. Instead evolutionists are left with a set of subjective evidences as they peer into an organism's development history.

Consider Berra's claim that the human embryo manifests the gills of its fish ancestor. Let us briefly review the biology involved. Fish use their gills for gas exchange. The main need for fish is obtaining sufficient oxygen from the water. The oxygen content in water is about twenty times less than that of air. Both air- and water-breathers rely on the process of diffusion to obtain their oxygen, but that process is about 300,000 times slower in water. Also, because water is denser than air, more energy is required to move the fluid over the gas-exchanging surface. On top of all this, warm water presents even more of a problem for fish, because its oxygen content is reduced while the fish metabolism is increased, thus requiring more oxygen. Fish overcome these obstacles with a sophisticated active gas transport system—gills—which transports oxygen from the water to the blood.

First, the many gill filaments that float off the gill arches provide a large surface area over which oxygen diffuses from the water into the bloodstream. Next, a near-continuous flow of water over the gills is created by a combination of positive pressure from the mouth cavity upstream of the gills and negative pressure from the opercular flaps downstream of the gills. The opercular flaps also serve to protect the vulnerable gills. Finally, the diffusion rate is maximized with unidirectional and opposing fluid flows. That is, the external water flow and the internal blood flow do not reverse direction (as with the air in our lungs, for example) but continue along in the same direction. Also, the water and blood move in opposite directions. This *countercurrent flow* produces a higher diffusion rate than if they flowed along together in the same direction.

The human embryo, on the other hand, obtains its oxygen (and nutrients) from its mother's bloodstream. It has no need for its own gas exchange system, and it certainly never has gills. It does have folds in the skin that superficially resemble gills, but they serve no function akin to a fish's gills. Berra admits that vertebrate embryos do not respire by means of gills, but he maintains that they "nevertheless pass through a gill-slit stage complete with aortic arches and a two-chambered heart, like those of fish."[37] But while human embryos do begin with a single-chambered heart that develops into a two-chambered heart, it later develops back into a single-chambered heart before redeveloping later to a two-, three-, and finally four-chambered heart. So we must ask: What is the justification for Berra's claim aside from subjective observations?

Or consider the claim that the three stages of mammalian kidney development are homologous to the adult kidneys of evolutionary ancestors. In fact, these three stages are not stages in the development of the final kidney organ; rather they are three different, successive, fully functional organs, only the last of which survives into adulthood. As Romer and Parsons explain:

> We see the development in the amniote embryo of three successive kidney structures: pronephros, mesonephros, and metanephros. It is often stated or implied that these three are distinct kidneys that have succeeded one another phylogenetically as they do embryologically. However, there is little reason to believe this. The differences are readily explainable on functional grounds.[38]

Of course, evolutionists could opt for an even weaker form of the biogenetic law, relieving embryological structures of the need of any functional similarity to the purported ancestor in favor of a mere similarity of appearance. But this would push the evidence that much further into the subjective realm, allowing each observer to invoke arbitrary criteria in assigning connections to ancestors.

Though nineteenth-century embryology seemed to promise us a backdoor peek at the history of evolution, we cannot be sure this is what we are really seeing. Darwin observed that the embryo is the animal in its less modified state, but his conclusion that it manifests the adult structures of related species remains unproven.

DNA Reveals Complexity, Not Evolution

Evolutionists also claim they find supporting evidence in molecular biology. In fact, it provides the ultimate homology, the universal genetic code that is used to encode information on the DNA molecule. Before we critique this claim, we need to briefly review how the code is used.

Put very simply, there are four different chemical building blocks called nucleotides that can be used in any order to construct the DNA strand. We can think of these nucleotides as letters: just as we arrange letters to form words and sentences, so too the nucleotides form a message encoded in the DNA strand. One minor difference is that in the DNA string there are a total of four letters in the alphabet, and all words have exactly three letters.

The information stored in the DNA strand is used when first a copy is made of it, and then, after some intermediate editing, the copy is passed over to a *ribosome*—structures in the cell that construct proteins by translating the message. Each nucleotide triplet in the DNA copy represents an amino acid according to the universal genetic code, and in the translation process the appropriate amino acids are lined up and glued together to form a protein, as specified by the information encoded in the triplets. But the ribosome cannot do this job alone. It has a whole population of tRNAs (transfer ribonucleic acids) at the ready. Each tRNA is armed on one side with an amino acid and on the other side with a nucleotide "reader." The key to the process is that the tRNA nucleotide reader recognizes the triplet that corresponds to the amino acid it carries. When it reads that triplet in the DNA copy, it proceeds to glue its amino acid onto the chain of amino acids that the ribosome is building.

But how do the tRNAs become armed with the right amino acid? It turns out that there are special proteins that assemble the two structures, tRNA plus amino acid, together. These special proteins have an inherent knowledge of the universal genetic code in the sense that they recognize and attach the tRNA and its corresponding amino acid—they know which amino acid goes with which nucleotide triplet. But, of course, these special proteins, like all proteins, are constructed by the very translation process we are describing here. They are constructed by the process that requires their prior existence. How can this be? The simple answer to this mystery is that the organism begins life with a complete and functional single cell—the zygote. The zygote already has the special proteins when it is first conceived. From there, more of the special proteins can be constructed as needed. This is one reason biologists refer to the cell as the basic unit of life.

But this simple answer only pushes the mystery down to a deeper level. If the zygote is fully functional to begin with, then how did the first zygote come about? Or, getting back to the DNA code, we can state the issue even more simply. The existence of a code implies that two distinct entities—the sender and receiver—must know the code *before* the message is sent. Therefore the existence of the DNA genetic code requires elaborate and coordinated sending and receiving machinery to be in the cell when a new individual is first conceived.

One might think that the twentieth century's discovery of the genetic code and the associated cellular machinery might have cast some doubt on the theory of evolution. For whereas earlier Darwinists might have hoped for simple beginnings, biology now knows that the cell not only is highly complex but also shows no signs of intermediate or abbreviated forms. In a letter Darwin speculated of a warm little pond with a protein compound ready to undergo more complex changes. Darwin's credulous acceptance of a spontaneous increase in complexity set the tone for evolution's response to the twentieth century's findings. The immense complexity of the cell, including the genetic code and DNA molecule, were seen not as a challenge to evolution but as supporting evidence, despite the fact that evolution could not explain how such complexity could have originated.

The problem is further complicated by the finding that the information encoded on the DNA strands can be overlapping.[39] That is, in a given segment two separate messages can be overlaid on each other. For example, in the word *evolution* we can find the word *love* by reading backwards. Overlapping DNA messages were generally considered not possible before their discovery in the 1970s, but now we must believe not only that random variations are the source of the meaningful information encoded on the DNA strand but that those variations also produced overlapping messages and the capability to read such messages.

Given the complexity of the cellular machinery and genetic code, it is not surprising that evolutionists do not have any detailed hypothesis about how it could have originated or evolved. Instead they have a wide variety of speculative ideas. Some evolutionists believe that the genetic code arose as a result of interactions with clay minerals. Others try to explain it as a result of nonenzymatic chemical reactions, and yet others have tried stereochemical approaches. An entirely different set of hypotheses holds that the genetic code arrived on earth from outer space, on meteors, comets, or spores driven by radiation pressure or even deliberately planted by extraterrestrial beings.[40]

In addition to the origin of the code, there are a variety of hypotheses about how the modern code could have evolved from a simpler code. Perhaps fewer amino acids were originally coded for, or perhaps the code distinguished between classes of amino acids rather than specific amino acids. Perhaps the alphabet was originally binary, or perhaps the words were only two letters long. Perhaps the original machinery was imprecise, so that a given gene did not always code for the same protein.[41]

One thing evolutionists do agree on is that there is a great deal of uncertainty about how the genetic code came about.[42] All the various hypotheses are grappling with the problem of finding a non-Darwinian mechanism for the genetic code. Because the code is chemically arbitrary, it holds no apparent competitive advantage over any other code. Swap in another code and things would work just as well, and therefore Darwin's law of natural selection is powerless to help explain the origin of the code.

To summarize the discussion so far: we have seen that the genetic code and associated machinery are highly complex, that there is a great deal of uncertainty about how it might have originated and evolved, and that Darwin's law of natural selection cannot be used to solve the problem. These difficulties show not only what a challenge evolutionists have but also that the genetic code's universality cannot be viewed as strong evidence for evolution, for the various hypotheses about how the genetic code could have originated and evolved are significantly different. Evolutionists are not at the stage where they are refining the minor details of a theory. Rather, the different hypotheses reveal fundamental differences of opinion.

It is natural for science to go through this stage in the early development of a theory. The problem here is that evolutionists are claiming the genetic code as evidence for their theory when the code's very existence remains unexplained. We have no idea how the genetic code originated; therefore we can hardly appeal to its existence as evidence for evolution.

There is yet another reason that the universality of the genetic code is not strong evidence for evolution. Simply put, the theory of evolution does not predict the genetic code to be universal (it does not, for that matter, predict the genetic code at all). In fact, leading evolutionists such as Francis Crick and Leslie Orgel are surprised that there aren't multiple codes in nature.

Consider how evolutionists would react if there were in fact multiple codes in nature. What if plants, animals, and bacteria all had different codes? Such a finding would not falsify evolution; rather, it would be incorporated into the theory. For if the code is arbitrary, why should there be just one? The blind process of evolution would explain why there are multiple codes. In fact, in 1979 certain minor variations in the code were found,[43] and evolutionists believe, not surprisingly, that the variations were caused by the continuing evolution of the universal genetic code.[44] Of course, it would not be a problem for such an explanation to be extended if it were the case that there were multiple codes. There is nothing wrong with a theory that is comfortable with different outcomes, but there is something wrong when one of those outcomes is then claimed as supporting evidence. If a theory can predict both A and not-A, then neither A nor not-A can be used as evidence for the theory. When it comes to the genetic code, evolution can accommodate a range of findings, but it cannot then use one of those findings as supporting evidence.

The universal genetic code is not the only homology claimed by evolutionists at the molecular level. Similar versions of the cellular machinery, such as the ribosomes, tRNAs, and proteins discussed above, are also found in all species. But while variations in the DNA code among different species are relatively rare, variations in the cellular machinery are common. Evolutionists explain these variations with speculative accounts of how such anomalies may have arisen. Speculation is sometimes necessary in science, but it also reveals a weakness in the use of the biochemical data as evidence for evolution. Consider the following two examples of how biochemical findings have forced evolutionists to resort to speculative explanations.

In the early days of microbiology, all cells were categorized into one of two groups. There were the larger, more complicated cells called eukaryotes and the smaller, simpler ones called prokaryotes. The eukaryotes had well-defined internal structures such as a nucleus, whereas the prokaryotes appeared to have no such degree of organization. It seemed obvious to evolutionists that the prokaryotes arose first in the history of life and that the eukaryotes appeared later as evolutionary descendants of the prokaryotes. But in the 1970s researchers began comparing the genetic material of prokaryotes and eukaryotes in detail. They made two interesting discoveries: first, there appeared to be a third category of cell type, and second, the three different types were sufficiently different that they could not have evolved from each other. The three categories were similar in many ways, but they also were sufficiently distinctive that no evolutionary relationships could exist between them.

Did life arise three times to produce three different lineages with all their similarities? No, evolutionists postulated that the three lineages must have evolved from a single progenitor.[45] A single progenitor evolving in three different directions would explain how the three lineages could have substantial similarities yet did not have any direct evolutionary relationship between them. But the problem with this explanation is there is no evidence for the progenitor. Postulating that it existed and evolved to produce the three different lineages is speculative and is mainly supported by the assumption that evolution is true.

The second example is really a continuation of the same story, for the next step was to piece together what the progenitor would have looked like by comparing the genetic differences and similarities of the three lineages. But the task became confusing due to the wide variety of genes between and among the three lineages. No clear picture of a simple progenitor emerged; instead, the only solution seemed to be a superprogenitor that already had most of the highly complex traits found in each of the three lineages. The superprogenitor would have been as complex as modern cells yet would have somehow arisen in a short time.

Again, evolutionists have proposed a speculative scheme to explain the problem. Perhaps the distribution of genes in the three lineages could have been arrived at if the progenitor was so rudimentary that genetic material was readily exchanged between cells in the same population. The process is roughly akin to what is known as *lateral gene transfer* in modern cells, but on a grander scale. The result would be that evolution would occur more between neighbors than between parents and offspring— horizontally rather than vertically.[46]

But this scheme is even more speculative than the preceding one. Not only is there no evidence for such an evolutionary process, but there is a wealth of missing detail. The point here is not that such schemes are

39

necessarily impossible but simply that they are speculative and demonstrate how the evidence from biochemistry fails to support evolution as overwhelmingly as many have claimed. What we have here is not independent evidence providing a new and unique confirmation of evolution but rather observations that are given an evolutionary interpretation.

Molecular Comparisons Do Not Require Evolution

The molecular phylogenies also fail to provide evidence for evolution as is often claimed. Recall that David Penny reconstructed the phylogeny for a group of eleven species, using five protein molecules. The proteins were used one at a time, independently of the other four, yet they suggested similar phylogenies. For Penny, Ridley, and others this provided "strong support" for the theory of evolution.[47]

In critiquing this approach, we should first note that though Penny found the trees to be "very similar," there were significant differences. For example, some of his trees show the dog relatively far from the human (nine species distant out of a possible ten), whereas others show the dog relatively close to the human (three species distant out of ten). The same is true for the mouse. Casual inspection of the trees reveals significant differences.

Penny obtained his trees by culling those that were most parsimonious—that is, he selected the trees that showed the least amount of evolutionary change to represent the history of life. The first problem is that Penny's method works perfectly fine on things we know did not come about via Darwinian evolution. For example, Penny's method would also claim that automobiles evolved from one another. Consider a group of vehicles, beginning with a small economy car and increasing in size to larger cars and to minivans and large-sized vans. One could quantify several aspects of the vehicle designs, such as tire size, steering mechanism, engine size, number of seats, and so forth. Presupposing the evolutionary paradigm and searching for parsimonious relationships, we would find that most of the design measures suggest the same relationship. The smaller vehicles have smaller tires, manual steering, smaller engines, and fewer seats. The larger vehicles have larger tires, power steering, larger engines, and more seats. In other words, the groupings suggested by the different design measures (tire size, steering mechanism, engine size, etc.) tend to be similar. But of course, the family of automobiles did not evolve from one another via random variations. The groupings of the design measures are a natural result of engineering and have nothing to do with Darwinian evolution. How then can Penny's results provide "strong support" for evolution?

Another problem with Penny's evidence comes from the general problem of theory falsification. Philosophers such as Willard Quine and Imre Lakatos have noted that the falsification of a theory is often more involved than simply finding a false prediction. The laws of logic state that if theory P predicts observation Q, then the failure to observe Q implies P is false. This form of reasoning is known as denying the consequent, or *modus tollens* (the way of denial). But in the real world things are rarely so straightforward. For example, the failure to observe Q could be interpreted as a failure in the experimental apparatus, or it could be sustained by appropriately modifying P rather than rejecting it altogether. Lakatos gives the example of a newly discovered planet that seems to violate Newton's laws of motion. Rather than reject the universality of Newtonian physics, the scientist could opt for all sorts of alternative explanations. Theories are not usually simple statements but a complex web of hypotheses and beliefs that can accommodate all sorts of adjustments before being abandoned.[48]

In the case of Penny's experiment, he assumed that the maximum parsimony model is a good predictor for molecular evolution. If the expected results had not been observed in the experiment, then the maximum parsimony model could have been replaced with another. For example, if the different trees suggested by the different proteins had not been similar, Penny could have concluded that lateral gene transfers were responsible for the discrepancies. On the other hand, Penny could have concluded that molecular evolution had been so rapid that information about the evolutionary histories had been lost. This is no mere hypothetical conjecture, for Penny used only five proteins out of thousands. Subsequent studies have shown that proteins often do not cooperate by suggesting similar trees.[49] Penny was aware of these limitations but ignored them when he concluded that "the existence of an evolutionary tree for these taxa is a falsifiable hypothesis." The evolutionary tree can be modeled using the maximum parsimony model or some other model, depending on which one best fits the data at hand. In order for the hypothesis to be falsifiable, evolutionists would have to commit themselves to a particular model.

More important, Penny claimed that his results provided "strong support" for the theory of evolution. If Penny had been given data from automobiles disguised as molecular data, he would have found evolutionary relationships. Furthermore, if he had been given random, uncorrelated data, he could have simply concluded that the maximum parsimony model was a bad assumption because the molecular evolution was too fast. Therefore the claim of providing strong support for the theory of evolution is unjustified.

Molecular Clock Is Not Evidence for Evolution

In addition to using molecular data to compute hypothetical phylogenies, evolutionists try to use molecular data as a clock. That is, whereas phylogenies purport to show the evolutionary relationships and branching points where species split off, evolutionists believe that molecular data can also be used to compute how many millions of years it has been since a pair of species split off from a common ancestor.

How reliable is the molecular clock? The evolution literature is full of instances where the molecular clock is apparently inaccurate. For example, it was found early on that different types of proteins must evolve at very different rates if there is indeed a molecular clock. Furthermore, it was found that the evolutionary rate of certain proteins must vary significantly over time and in different species.[50] The snake cytochrome c protein, for example, must have evolved several times faster than in other species.[51] It was also found that the molecular clock doesn't seem to work very well for bacteria. The molecular data make closely related bacteria look like distant relatives—as different as insects are from mammals, for example.[52] It was also found that the relaxin protein was anomalous when compared across different species. The pig, for example, was found to be more closely related to a shark than to a rodent.[53] "The conclusion to be drawn from the relaxin sequence data," wrote one researcher, "is that they do not fit the evolutionary clock model."[54]

Furthermore, in order to fit the data to the molecular clock hypothesis, one must imagine that different regions of the genome evolve at different rates for a species and that the same region evolves at different rates in different species.[55] Many instances have been discovered where the data do not seem to fit the molecular clock model, and some researchers wonder if this is where the real message is: "It seems disconcerting that many exceptions exist to the orderly progression of species as determined by molecular homologies; so many in fact that I think the exception, the quirks, may carry the more important message."[56]

There are, to be sure, explanations for the various anomalies. In fact, evolutionists have many possible reasons that data may not fit the molecular clock. For example, the molecular clock model depends on the species population size and likely on its life span—the generation time effect. Also, varying DNA replication accuracies in different species may affect the clock; or could it be possible that the protein under study has somehow changed its function in its evolutionary history? On the other hand, perhaps a horizontal gene transfer has taken place at some point in the organism's history, or perhaps so many mutations have occurred that the picture has become blurred. Perhaps molecular evolution undergoes elevated rates during periods of adaptive radiation, or maybe slightly deleterious mutants were incorporated during population bottlenecks.

It might seem that, given this battery of explanatory devices, there would be no observation that evolutionists could not explain. But there are, in fact, some cases that remain difficult to explain. For example, the serum albumin gene family shows significant deviations from clocklike evolution. Researchers who investigated these genes concluded that the molecular clock "is subject to the same vagaries as the rest of biology. Models are only models; they are only as good as the underlying assumptions. And if the number of assumptions (unknowns) is greater than the number of equations, a rigorous solution is but an illusion. This seems to be the case with the molecular clock."[57]

Other erratic molecular clocks include those based on the superoxide dismutase (SOD) and the glycerol–3-phosphate dehydrogenase (GPDH) proteins. On the one hand, SOD unexpectedly shows much greater variation between similar types of fruit flies than between very different organisms such as animals and plants. On the other hand, GPDH shows a more or less reverse trend for the same species. As one eminent scientist concluded, GPDH and SOD taken together leave us "with no predictive power and no clock proper."[58]

Another problem with the molecular clock hypothesis is that the clock must be calibrated before it can be used. Evolutionists use fossil data to calibrate the clock. Fossils are used to reconstruct a hypothetical evolutionary tree—a phylogeny—including the geological time since particular speciation events. The molecular differences between species are calibrated to those speciation events. The molecular clock is then used to measure the time since other speciation events. In many cases the molecular clock conflicts with the fossil data. That is, once the clock is calibrated using one part of the fossil record, it conflicts with another part of the fossil record. The molecular and fossil data often do not give similar results. This can be explained, for example, by pointing to possible errors in the fossil record. But the problem in all this is that the molecular clock is being calibrated with the same sort of data that the clock then contradicts. As one researcher put it: "Even if one makes the bold assumption that molecular clock models have little error, there seems little objective reason for accepting a few fossil dates used in calibrations and rejecting as unreliable the much more numerous fossil dates that contradict the resultant molecular estimates."[59]

There are plenty of anomalies that challenge the molecular clock hypothesis, and it has been controversial practically since it was proposed. While evolutionists continue to employ the hypothesis, its validity remains in question. No doubt for those who believe in evolution the molecular comparisons are another piece of information that must be considered, but it is a considerable stretch to claim the molecular clock as evidence for evolution.

Metaphysical Arguments

How is it that evolutionists can rely so heavily on the argument from homology, an argument that raises more questions than it answers? How can Darwin admit that his theory can "only to a certain extent" explain homologies and yet at the same time proclaim that, based on homology, he would "without hesitation adopt [evolution], even if it were unsupported by other facts or arguments"?[60] The answer may lie in the metaphysical interpretation of homologies that evolutionists often use.

God Would Never Repeat a Pattern

The genetic code and the DNA molecule are often cited as homologies that provide strong evidence for evolution. As we have seen, these claims raise many scientific questions, but there is a nonscientific interpretation of this evidence to which evolutionists often appeal. They see the genetic code and DNA molecule as evidence *against* the doctrine of divine creation. For example, Ridley claims that whereas the genetic code is preserved across species, it would not be if the species had been created independently.[61] Apparently Ridley believes that if there is a Creator, then he is obliged to use different genetic codes for the different species. Similarly, Berra claims that the theory of evolution is "the only reasonable explanation" for the fact that virtually all organisms carry their genetic information in the DNA molecule.[62] In other words, this homology is not positive evidence in favor of evolution but rather negative evidence against the competition.

In the protein comparisons Penny's null hypothesis was random trees. It mattered little that the "similar" trees actually had significant differences. The key finding was that the parsimonious trees were *far from random*, and the unspoken premise is that the doctrine of divine creation predicts random trees because each species is created independently. The experiment did not so much prove evolution as it disproved the evolutionist's view of creation. "Clearly," Penny proclaimed, "we can reject any idea that the trees from the different sequences are independent."

Ridley agrees and supplies his own version of this metaphysical claim. "If the 11 species had independent origins, there is no reason why their homologies should be correlated," and "if they were independently created, it would be very puzzling if they showed systematic, hierarchical similarity in functionally unrelated characteristics." Ridley does admit other explanations are possible, but they all are really just equivalent to the original conclusion against independent creation. "Of course, some innocent explanation might be found for any such correlations. . . . Maybe they could all be explained by class of owner, or region, or common architects. But that is another matter; it is just to say that the buildings were not really independently created."[63] It seems that for Ridley the notion of a "common architect" does not support divine creation. Of course, Ridley is entitled to whatever metaphysical view of God and the world he prefers, but he is using that view to support the theory of evolution, and this is the point.

Penny's use of the five different proteins as evidence for evolution really amounts to an argument from classification. For centuries naturalists have observed that species can be classified into groups—species tend to fall into the same groupings based on different anatomical features. Species with similar bones are likely also to have similar bladders or brains. There are plenty of exceptions, but groups definitely exist. If one presupposes evolution, then these groupings can be used to derive evolutionary trees representing evolutionary relationships between the species, but it is difficult to make the argument that the groupings themselves are scientific evidence for evolution. Some evolutionists have tried, but they end up smuggling religious premises into the argument. For example, after discussing the groupings of species, Darwin concluded that "these are strange relations on the view that each species was independently created" and that it was "utterly inexplicable on the theory of creation."[64] Evolutionists since Darwin who have tried to make an argument from classification have also relied on metaphysics (see chapter 5).

Biochemistry can also be used to classify species. Penny's use of protein sequences is one of several possible ways to compare species at the molecular level. It is more quantitative than traditional techniques based on visible features, but it is conceptually similar. And it is not surprising that, like the arguments based on the visible features, Penny's and Ridley's arguments based on biochemistry are ultimately metaphysical.

Evolutionist Douglas Futuyma is also impressed by the variations in protein sequences and the phylogenetic relationships they suggest. For why, when we find the same molecules in different species, are they not identical? Why are the hemoglobin molecules slightly different in different species? For Futuyma this does not square with creation. "A creationist," he argues, "might suppose that God would provide the same molecule to serve the same function, but a biologist would never expect evolution to follow exactly the same path twice."[65] Futuyma and the others seem to be unaware that their argument hinges on their view of God. They issue their polemics as though they were scientific findings.

Likewise for Darwin the real strength of his homology argument came when a nonscientific interpretation was attached. In Darwin's day the doctrine of divine creation stated that God created the species independently—species were supposed to have been uniquely designed, and common patterns found among different species reflected God's plan. Perhaps they were required for good engineering design. But nineteenth-century naturalists were finding patterns that did not seem to reflect a thoughtful plan or good engineering design. Nature seemed to make use of the same design over and over, even though the need was quite different. Creation was not fulfilling the naturalist's expectations, and for Darwin this was very curious:

> What can be more curious than that the hand of a man, formed for grasping, that of a mole for digging, the leg of the horse, the paddle of the porpoise, and the wing of the bat, should all be constructed on the same pattern, and should include similar bones, in the same relative positions? How curious it is, to give a subordinate though striking instance, that the hind-feet of the kangaroo, which are so well fitted for bounding over the open plains,—those of the climbing, leaf eating koala, equally well fitted for grasping the branches of trees,—those of the ground-dwelling, insect or root-eating, bandicoots,—and those of some other Australian marsupials,—should all be constructed on the same extraordinary type, namely with the bones of the second and third digits extremely slender and enveloped within the same skin, so that they appear like a single toe furnished with two claws. Notwithstanding this similarity of pattern, it is obvious that the hind feet of these several animals are used for as widely different purposes as it is possible to conceive. The case is rendered all the more striking by the American opossums, which follow nearly the same habits of life as some of their Australian relatives, having feet constructed on the ordinary plan.[66]

In this example Darwin points out that there are common patterns among different species—common patterns that do not seem to reflect good engineering design. The strength of the argument lies in its implicit rebuke of divine creation, for why would God have designed such a mundane world? Why would God use the same pattern for different uses, and as Darwin observed, sometimes even within one species?

> How inexplicable are the cases of serial homologies on the ordinary view of creation! Why should the brain be enclosed in a box composed of such numerous and such extraordinary shaped pieces of bone, apparently representing vertebrae? . . . Why should similar bones have been created to form the wing and the leg of a bat, used as they are for such totally different purposes, namely flying and walking? Why should one crustacean, which has an extremely complex mouth formed of many parts, conse-

quently always have fewer legs; or conversely, those with many legs have simpler mouths? Why should the sepals, petals, stamens, and pistils, in each flower, though fitted for such distinct purposes, be all constructed on the same pattern?[67]

Though Darwin did not know how the design of the crustacean or the flower could have been improved, he believed there must have been a better way and that God should have used it. God, according to Darwin, would not have made the brain or the bat that we find in nature, though he had little idea about how they actually worked. One could just as easily argue that the Creator used the patterns found in homologous structures so that scientists could more easily analyze his creations and figure out how biology works. On the other hand, one could argue that the imperfections of nature that the homologies reveal are a manifestation of the burden of sin upon the world. Such ideas are no more religious than evolution's notion of a restricted god.

Evolution Proved by the Process of Elimination

Negative theology was a consistent theme for Darwin, and it remains popular with today's evolutionists. The theory of evolution is true not because species obviously evolved from each other but because of the failure to reconcile God and nature. Darwin studied orchids in detail and found underlying patterns. The orchids seemed to have been made of spare parts rather than individually created. For Darwin and modern evolutionists this argues for evolution because it argues against the possibility of divine creation. Evolutionist Stephen Jay Gould sums up the argument as follows:

> Orchids manufacture their intricate devices from the common components of ordinary flowers, parts usually fitted for very different functions. If God had designed a beautiful machine to reflect his wisdom and power, surely he would not have used a collection of parts generally fashioned for other purposes. Orchids were not made by an ideal engineer; they are jury-rigged from a limited set of available components. Thus, they must have evolved from ordinary flowers.[68]

Notice how easy it is to go from a religious premise to a scientific-sounding conclusion. The theory of evolution is confirmed not by a successful prediction but by the argument that God would never do such a thing. For evolutionist Kenneth R. Miller, pseudogenes, which he believes are nonfunctional, reveal a designer who "made serious errors, wasting millions of bases of DNA on a blueprint full of junk and scribbles."[69]

47

Scientific theories make predictions, and if a prediction turns out to be false, then the theory is proven wrong. But a correct prediction does not mean that the theory is true—theories can always be falsified by subsequent experiments. Therefore in science it is not clear how theories can be declared to be true on the basis of experiments. But if opposing theories can be falsified, then perhaps a theory can be proven true by the process of elimination. As Gould put it:

> Odd arrangements and funny solutions are the proof of evolution—paths that a sensible God would never tread but that a natural process, constrained by history, follows perforce. No one understood this better than Darwin. Ernst Mayr has shown how Darwin, in defending evolution, consistently turned to organic parts and geographic distributions that make the least sense.[70]

In other words, evolution is true because divine creation is false. Likewise, "why," asks Ridley, "if whales originated independently of other tetrapods, should whales use bones that are adapted for limb articulation in order to support the reproductive organs? If they were truly independent, some other support would be used."[71] Of course it is evolutionary conjecture that the whale bones are adapted for limb articulation, but Ridley's main point is that God would not have created whales like this. Similarly, Gould asks: "Why does our body, from the bones of our back to the musculature of our belly, display the vestiges of an arrangement better suited for quadrupedal life if we aren't the descendants of four-footed creatures?"[72] And for Berra "the passage of a fishlike stage by the embryos of the higher vertebrates is not explained by creation, but is readily accounted for as an evolutionary relic."[73]

Futuyma also gives a series of such rejections of any possibility of God's hand in nature. Here are four paragraphs quoted in sequence, but not continuous in the original:

> If God had equipped very different organisms for similar ways of life, there is no reason why He should not have provided them with identical structures, but in fact the similarities are always superficial.

> The facts of embryology, the study of development, also make little sense except in the light of evolution. Why should species that ultimately develop adaptations for utterly different ways of life be nearly indistinguishable in their early stages? How does God's plan for humans and sharks require them to have almost identical embryos?

> Take any major group of animals, and the poverty of imagination that must be ascribed to a Creator becomes evident.

When we compare the anatomies of various plants or animals, we find similarities and differences where we should least expect a Creator to have supplied them.[74]

These quotes from a number of modern evolutionists show how prevalent and persistent is Darwin's metaphysical argument. Behind this argument about why the patterns in biology prove evolution lurks an enormous metaphysical presupposition about God and creation. If God made the species, then they must fulfill our expectations of uniqueness and good engineering design. We might say that God was supposed to have optimized the design of each species. Evolutionists have no scientific justification for these expectations, for they did not come from science. They are part of a personal religious belief and as such are not amenable to scientific debate. In fact, evolutionists rely on a rather narrow metaphysical target for their attacks on creation. The evolutionist's notion of God and divine creation is, for many people, just a straw man—an overly simplified metaphysic that conveniently supports their views.

With evolution, metaphysical arguments that the fauna and flora fit into the Creator's plan became unacceptable, but counter metaphysical arguments that the fauna and flora *could not* fit into the Creator's plan were acceptable. Darwin did not liberate biology from metaphysical thought as is sometimes claimed—he merely switched the metaphysics. What was right is now wrong, and vice versa. Darwin's version of homology is no more *scientific* than the pre-Darwinian version, for when evolutionists use a metaphysical interpretation of homology as an argument against creation, they move from the scientific to the metaphysical realm.

3

Small-Scale Evolution

Evolution calls for a massive amount of change to have occurred as evolution shaped the millions of diverse species that populate the earth. Evolutionists are not sure how all this change came about, but one idea is that it comes from the accumulation of a long sequence of small changes. Therefore, to the extent that this idea can be verified, small-scale changes that we observe taking place in today's plant and animal populations serve as evidence for evolution.

This line of evidence raises many doubts about evolution. Nonetheless, evolutionists routinely appeal to small-scale change, sometimes even claiming that it provides virtual proof of their theory. This chapter separates the scientific and unscientific small-scale change arguments by first looking at the scientific arguments for and against evolution and then looking at how the strength of the evidence lies in its metaphysical interpretation.

The Evidence

On his voyage aboard the HMS *Beagle,* Darwin saw closely allied species in the wild. The most famous example was the wide variety of finches Darwin observed on the Galápagos Islands six hundred miles off the coast of South America. The finches, as well as the mockingbirds and tortoises Darwin observed, varied distinctly from island to island. Some finches lived in coastal areas on the ground, others lived in forest trees, yet others lived in bushes. And the diets of these varieties varied considerably. One of the species ate buds and fruit, another prickly pear; others ate seeds, and others were insectivores. One of the insectivores even used a twig to fish out insects from crevices in tree bark. Nicholas Lawson, the vice governor who entertained Darwin over dinner, claimed that so distinct were the tortoises from island to island that given the tortoise shell he could identify the island of origin.

Much of this evidence lay muddled up in Darwin's head, but eventually, back in London, he would theorize that the different finches, mockingbirds, and tortoises were indeed different *species* that nature had bred in the wild, using individual variations as the initial raw material. In other words, Darwin believed that these were new species that had arisen naturally. Darwin would argue that the different species resulted from an accumulation of small-scale change. Nature, it seemed, had created new species, albeit only slightly different ones. Darwin would next argue that these slight variations could be extrapolated to large changes. If nature could use small-scale change to break the species barrier, could this not lead to even greater change given enough time?

There are also populations that change gradually over a broad geographic region. At the opposite ends of the region, individuals are so different that they seem to represent two different species. From one end of the region to the other, nature seems to have evolved a new species. In one such case, gulls are found in a range that extends around the Northern Hemisphere of the globe. Two species of gull meet in northern Europe: the herring gull and the lesser black-backed gull. The herring gull in North America is similar to its European cousin, but in Asia the gull begins to take on the appearance of Europe's other gull, the lesser black-backed gull. The trend continues across Asia until the gull becomes the lesser black-backed gull in northern Europe. This ring presents a varying population around the globe except in northern Europe, where the ring meets and two species of gull overlap geographically but do not interbreed. The point is that small-scale change adds up to two different types of gull. It seems to be an example of gradual evolution leading to two different species. Instead of distinctly isolated and fixed types, populations sometimes vary slowly and continuously.

In the cases of the finches and gulls, the hypothesized speciations took place long before scientists were around to observe the changes. But there are small-scale changes that we can observe. A favorite example for evolutionists is the melanism brought about in moths by industrial pollution.

Other examples of small-scale change include insect populations that become resistant to pesticides and bacteria populations that modify their metabolism to adapt to strange environments.[1] Substantial evolutionary change has also been brought about by breeding efforts. Such *artificial selection* has developed a wide variety of domesticated animals and plants. In the domestic dog, *Canis familiaris,* the various breeding lines are so different that they would be judged to be different species if found in the wild. Other such examples range from dairy cows to one of Darwin's favorites, pigeons. Such human-directed evolution has also produced the bulk of our agricultural crops and horticultural products. Most of today's popular orchids, irises, tulips, and dahlias have been artificially developed.

Problems with the Evidence

This brief review shows that small-scale evolutionary change is real. But small-scale change raises as many questions about evolution as it answers.

Biological Modification Is Limited

Darwin's book was entitled *The Origin of Species,* but he never did actually explain how species originate—the process of speciation. This problem remains unresolved 140 years later. There is little in the way of detailed or convincing explanation of how the required genetic rearrangements would take place. "In spite of all the advances in genetics," wrote Ernst Mayr in 1988, "we are still almost entirely ignorant as to what happens genetically during speciation."[2]

One of the thorny obstacles regarding speciation is the fact that a population's capacity to change seems limited. Instead of small changes accumulating and resulting in large changes, the small changes appear to be bounded. Darwin was well aware of this problem. He bred pigeons and made it his business to understand state-of-the-art animal husbandry and breeding. And the state of the art, then and now, is a story of change within limits. One can successfully bring out all sorts of features in a population of pigeons, dogs, horses, and the like, but there seem to be definite limits—one cannot modify pigeons to become dogs or horses. Furthermore, specialized breeds produced by artificial selection are not more fit for survival and reproduction. Specialization and differentiation can be achieved, but it is at the expense of the overall fitness of the individual. One thinks of the purebred dog that is less hardy than the mutt.

The idea that biological change is limited had been considered before Darwin. The great German writer-philosopher Johann Goethe (1749–1832) became interested in nature later in his career. He contended, along with the French naturalist Étienne Geoffroy St.-Hilaire (1772–1844), that there is in nature a law of compensation that limits biological variation. This doctrine was a problem for Darwin, for it held that species are anchored to their design—they can drift about a bit, but they cannot continue to evolve to form a new species.

Darwin could not simply reject the notion of limited change, for it was known to be true from breeding experiments. So Darwin argued that it did not apply to organisms in their natural environments. "With species in a state of nature," Darwin argued, "it can hardly be maintained that the law [of compensation] is of universal application," even though "many good observers"[3] believed it to be true. Darwin then cited two particular organisms that he claimed supported his position. Most readers were probably impressed with the wealth of biological details he presented in his argument, but as we shall see, his logic was a simple case of circular reasoning. Darwin argued that the unique features of these two organisms show that the law of compensation does not hold in the wild, but his unspoken premise was that the unique features *evolved*. Darwin presupposed the truth of evolution in order to find evidence for evolution.

Let's have a look at Darwin's argument. Cirripedia, an order in the class crustacea, are sessile and highly distinctive barnacles upon which Darwin had completed a major systematic study. Over against Goethe, Geoffroy, and his own breeding experience, Darwin argued that variation can be extrapolated by evolution to larger changes because such larger changes were evident in two genera named *Ibla* and *Proteolepas:*

> If under changed conditions of life a structure, before useful, becomes less useful, its diminution will be favoured, for it will profit the individual not to have its nutriment wasted in building up a useless structure. I can thus only understand a fact with which I was much struck when examining cirripedes, and of which many analogous instances could be given: namely, that when a cirripede is parasitic within another cirripede and is thus protected, it loses more or less completely its own shell or cara-

pace. This is the case with the male Ibla, and in a truly extraordinary manner with the Proteolepas: for the carapace in all other cirripedes consists of the three highly-important anterior segments of the head enormously developed, and furnished with great nerves and muscles; but in the parasitic and protected Proteolepas, the whole anterior part of the head is reduced to the merest rudiment attached to the bases of the prehensile antennae. Now the saving of a large and complex structure, when rendered superfluous, would be a decided advantage to each successive individual of the species; for in the struggle for life to which every animal is exposed, each would have a better chance of supporting itself, by less nutriment being wasted.

Thus, I believe, natural selection will tend in the long run to reduce any part of the organisation, as soon as it becomes, through changed habits, superfluous, without by any means causing some other part to be largely developed in a corresponding degree. And, conversely, that natural selection may perfectly well succeed in largely developing an organ without requiring as a necessary compensation the reduction of some adjoining part.[4]

Here Darwin is not simply applying his theory to a set of observations. He is not illustrating how evolution might have formed these species. Rather, Darwin is building an argument for evolution. He is trying to show how species are not limited in how much they can change. But in constructing his argument, he presupposes that the organisms had evolved. Darwin was unquestionably an expert on barnacles, and he could describe in detail examples of highly modified species that most people have never even heard of. But Darwin simply begs the question when he says that the distinctive structures of *Ibla* and *Proteolepas* evolved and therefore variation is not limited in the wild.

To summarize these ideas: First, many, including Darwin, observed that there is a limit on how much a species can be modified. If true, this would be a problem for the theory of evolution. Darwin argued that species modification is really limited only for domesticated species and that species in the wild are not so constrained. To support this claim, Darwin cited the *Ibla* and *Proteolepas*. These organisms are significantly different from their cousin organisms. Darwin argued that the differences were caused by natural selection acting on variation— that is, the differences were the result of evolution. Therefore in this case, Darwin concluded, the modification was significant—it was not limited.

Clearly, in order for Darwin to arrive at this conclusion he had to use evolution as a premise. He assumed evolution is true in order to argue for evolution—he begged the question.

Darwin presupposed the truth of evolution to argue that significant change is possible in the wild, but this still left unanswered just *how* evolution could achieve such great change. If breeders found that change had limits, how did evolution produce it so copiously? Once again Darwin solved the puzzle with a thought experiment. He argued that natural selection can produce all sorts of change that had eluded the breeders. One might suppose the breeder to be at an advantage over nature, as an intelligent guiding force with direct control over breeding activities. Not so, according to Darwin, for the breeder acts only on visible and external characters, while natural selection cuts to the heart of the problem. By acting only via survival and ultimately reproduction, natural selection acts on "every internal organ, on every shade of constitutional difference, on the whole machinery of life. Man selects only for his own good: Nature only for that of the being which she tends."[5] Darwin triumphantly concluded: "As man can produce, and certainly has produced, a great result by his methodical and unconscious means of selection, what may not natural selection effect?"[6]

Darwin is trying to show us how wrong our intuitions can be. Natural selection, it seems, is all the more efficient because it is not limited to certain traits as breeders are. This thought experiment is too speculative to be very compelling. Darwin did the best he could to elevate the powers of natural selection over the breeder's artificial selection, but in the end we are left with a mere possibility.

Small-Scale Change Often Reverts

Early in the twentieth century, a new theory for how small-scale change could extrapolate to large change was formulated. A contemporary of Darwin, Gregor Mendel (1822–1884), had made the first discoveries in what would become modern genetics. In 1865, just six years after Darwin published *Origin* and while he was still working on revisions, Mendel published a single paper on the laws of heredity. He was an Austrian monk who taught science and math and had performed years of experiments with the common pea plant, *Pisum sativum*. Mendel's work went unnoticed for many years and was finally rediscovered at the turn of the century.

It was not an easy task, but Mendel's findings were eventually merged with the theory of evolution. The upshot was that normal biological variation arises from different gene combinations. Breeders achieve their results by manipulating these combinations, but all the while they are operating within a given gene pool. This is why they come up against limits to the amount of change they can produce. The way to produce greater change would be to use new genes not present in the existing pool, and the way to produce new genes is via mutation.

Mendel was able to infer the fundamental principles of genetics because he had fortuitously studied traits that were controlled by uncomplicated genetic mechanisms. Later in the twentieth century scientists found that genome architecture and dynamics were much more complicated than Mendel's initial findings suggested. In any case, regardless of what mechanism is posited, the key tenet of evolutionary genetics remains that biological variation is unguided.

The fusion of Mendelian genetics with Darwin's evolution seemed to lend great credence to the latter. Before Mendelian genetics came to the fore, it was commonly thought that a continuous spectrum of traits is formed by the blending of parental traits, but this notion was subject to objections that it could not provide the kind of variation that evolution required. Mendelian genetics, on the other hand, called for a discrete spectrum of traits, which could support evolution's required variation. The objections about blended inheritance could be discarded; now there was an explanation for genetic variation and for how new species could potentially arise.

Great faith was placed in the power of randomly produced genes to provide the raw material upon which natural selection works. A theme in Darwinism is randomness, and since mutations are essentially random events, they seemed to fit perfectly with evolution. Researchers set about trying to induce mutations in the laboratory to see what new species they could create, but this area of work has not fulfilled early expectations. Viruses and bacteria can change significantly (although they are still viruses and bacteria), but in multicellular organisms mutations are almost universally not beneficial and there is a resistance to change.

Biological organisms have a remarkable ability to accommodate environmental changes imposed on them. Amoebae move toward food and away from extreme temperatures; trees grow "shade leaves" on their lower branches that have a higher chlorophyll concentration, and higher leaves exposed to direct sunlight may be thicker so they don't overheat; both fauna and flora are endowed with "biological clocks" to help respond to daily and annual variations in the environment; the blood's oxygen-binding characteristics change when animals acclimate to high altitude—the Bohr effect; muscles shiver in cold temperatures to convert chemical energy into heat. This ability of organisms to maintain internal equilibrium in the presence of external environmental change is referred to as *homeostasis*. This ability seems to be inherent in the genotype as well. Geneticist I. M. Lerner called it *genetic homeostasis*.[7]

As we saw earlier, modifications induced by breeders can hardly be called improvements to an organism's fitness. In genetic experiments with the common fruit fly, *Drosophila,* artificial selection for extreme traits usually results in sterility. And when purebred strains are permitted to return to natural breeding, human-directed modifications tend to dissipate. In one fruit fly experiment, a harmful mutation was naturally diluted even though it did not slow reproduction in the artificial laboratory environment.[8] One may wonder what would happen with a favorable mutation in the laboratory. We may never know because such mutations are difficult to find. Mutations have been induced in *Drosophila* using a variety of means, and they are not beneficial.

There seems to be an internal balance of pressures—a restoring action in the genetic system—and it is not clear that small-scale change can actually lead to the large-scale change required by evolution. This is ironic, since nature provides us with seemingly endless examples of variety. Species fill every niche and come in all shapes and sizes, yet there seems to be a barrier that inhibits the movement of any one species too far from its own design.

Unjustified Extrapolation

The discussion so far has shown that from Darwin's time to today biological modification has been found to be limited. The precise nature of this limit is not well understood, and in fact no hard limit has been identified. It is not that extrapolation of small-scale change to large changes is disproved; rather, it is not well supported. No one has yet proven that small-scale change does not extrapolate to large-scale evolution, but this does not mean that small change *does* add up to large change. But because no hard limit to modification can be found, evolutionists feel free to use small-scale change as evidence for evolution. Darwin, as we saw, had no qualms about making such an extrapolation, and today Mark Ridley appeals to the principle of uniformitarianism to justify such an extrapolation:

> It is possible to imagine, by extrapolation, that if the small-scale processes we have seen were continued over a long enough period they could have produced the modern variety of life. The reasoning principle is called *uniformitarianism.* In a modest sense, uniformitarianism means merely that processes seen by humans to operate could also have operated when humans were not watching; but it also refers to the more controversial claim that processes operating in the present can account, by extrapolation over long periods, for the evolution of the earth and of life. This principle is not peculiar to evolution. It is used in all historical geology. When the persistent action of river erosion is used to explain

the excavation of deep canyons, the reasoning principle again is uniformitarianism.[9]

We must agree with Ridley that such application of uniformitarianism is "possible to imagine," but a *possibility* does not count as a *probability*. The use of small-scale change as evidence for large-scale change is based on speculation.

The argument that small-scale change can be extrapolated to large-scale change relies on a strong distinction between what evolutionists call microevolution and macroevolution. Most of the small-scale change we have been looking at in this chapter is considered microevolution. When a change is great enough to cross the species barrier, then evolutionists place it in the macroevolution category.

There is no single objective definition of the species barrier. Biologists in fact use a half-dozen or so definitions depending on their purpose. One may define a species based on physical similarity, the ability to interbreed, mate recognition, niche adaptation, or purported evolutionary lineage. But even these definitions are subject to interpretation, and what may be a different species to one investigator is the same species to another. Nonetheless, when small-scale change is seen to cross the species barrier, evolutionists claim it as an example of macroevolution.

Macroevolution includes all the rest of the supposed evolutionary change: fishes becoming amphibians, amphibians becoming reptiles, reptiles becoming mammals, and so forth. Hence, so the argument goes, if one can prove one type of macroevolutionary change, then all types of macroevolution become possible. Obviously it is questionable to use such minor variation such as coloration changes in gulls to justify mammals arising from reptiles.

The Existence Problem

Mendel discovered the foundation of modern genetics, and the twentieth century's revolution in molecular biology has filled in the details. We now know that the molecular mechanisms that produce genetic variation are incredibly complex. Whereas early evolutionists might have envisioned a simple sort of random perturbing force, we have discovered a highly intricate Mendelian machine behind variation. These discoveries have been hailed as great victories for evolution, but there seems to remain a question that is rather fundamental: Where did the mechanism for variation come from?

This is sometimes called the *existence problem*. Evolution relies on the preexistence of biological variation without understanding where it came from. We now know how variation comes about but not how the machine

59

behind it came about. The shortcoming is particularly awkward because evolution proposes to tell us not only how variation is used but how all life came about—and its answer is, by unguided natural forces. But when we come to the Mendelian machine of variation we must ask how it was that evolution produced such a fine-tuned machine that is, in turn, supposed to be the engine for evolution itself.

Here we find an element of serendipity in evolutionary theory. Without variation, natural selection was powerless to work, yet somehow a source of variation arose. Speculation is alive and well about how this might have happened; for example, perhaps a simple source of variation somehow arose first and later evolved and gained complexity. This is, of course, possible, but we have no evidence for it. The point here is not that small-scale change could not have extrapolated to large changes, but simply that it is unreasonable for small-scale change to serve as strong evidence for evolution.

It is not that evolutionists have no argument from small-scale changes. They certainly do, but the argument brings along with it a multitude of unaccounted-for difficulties. It is interesting evidence, but it is also problematic. Evolutionist Niles Eldredge notes: "Small-scale evolutionary change can and does take place in short intervals of time—a circumstance consistent with the general notion of evolution."[10] Yes, the evidence is consistent with the general notion of evolution, but it's usually possible to find evidence to support even theories that we know are false. Every level pasture is evidence for a flat earth, but that doesn't mean the earth is flat. Moths may change color, and the beaks of birds may change size, but this does not mean that reptiles changed into birds.

At first glance, small-scale changes may seem to support evolution. But this could be a case of false appearances, for there are several difficulties that must be considered. Most obviously, the relationship between small-scale change and large-scale change is an open question even among those who believe in evolution.[11] And the small-scale changes arise from complex biological mechanisms whose existence evolution does not explain.

Despite these problems, Eldredge's sober assessment of the evidence often gives way to unwarranted confidence. Microevolutionary change is routinely used as one of the pillars establishing evolution as a fact. According to Ernst Mayr, "Evolutionary change is also simply a fact owing to the changes in the content of gene pools from generation to generation."[12] Likewise, Isaac Asimov claims that the peppered moth example proves evolution.[13] Steve Jones writes that the changes observed in HIV (the human immunodeficiency virus) contain Darwin's "entire argument."[14] And according to science writer Jonathan Weiner, the changes in the beaks of birds show us "Darwin's process in action."[15]

60

There is a disparity between the quality of this evidence and the high claims of evolutionists. How is it that evolutionists can be so confident? One reason may be that another interpretation of this evidence supports evolution indirectly by arguing against the doctrine of creation.

Metaphysical Arguments

In order to understand the metaphysical interpretation of the small-scale change evidence, we must briefly recall the Swedish botanist Carl von Linne, or Linnaeus (1707–1778). Linnaeus was one of the greatest naturalists of the eighteenth century. He is most famous for the universally accepted hierarchical method of classification that today bears his name. But Linnaeus's popularity and influence extended beyond classification. One of his most famous and influential beliefs, which was linked to his classification system, was the fixity of species. His proclamation *"Nullae species novae"*—no new species—idealized species as perfect forms created by the wise Creator. This doctrine reflected Linnaeus's religious beliefs about God and creation. He believed that creation is like a big museum full of proofs of God's wisdom and power, and the naturalist's purpose is to discover these evidences. Linnaeus attempted to rationalize creation, and he even constructed his own creation theory about how species populated the earth.[16]

These religious beliefs were typical of the modern trend. Put simply, modernism was tending toward a view of God as benevolent Creator. God was a wise craftsman who was manifest in his creation, which was full of beauty and harmony. For Linnaeus this meant that species were perfect and unchanging.

The Downfall of Fixity of Species

Linnaeus's fixity of species concept could accommodate an old earth with multiple creation events or successive revolutions. It could even accommodate extinctions. But it could not survive if science were to find that new species are routinely created by unguided natural forces.

Linnaeus was troubled when he discovered hybrids—species produced by the crossing of two related species—and he subsequently softened his doctrine of fixity of species. But this was inconsequential: his system with its conception of species became deeply rooted, and the nineteenth century began with the notion of *species as immutable* still strongly in place.

This notion was increasingly being challenged, but it was nonetheless a major obstacle for Darwin to overcome. It was therefore highly significant when Darwin became persuaded that related populations of birds he observed at the Galápagos islands were actually different species. If there was the slightest foundation for this idea, Darwin had written in a famous notebook entry, it "would undermine the stability of species."

The birds did not suddenly reveal to Darwin how fishes could change to amphibians, or how amphibians could change to reptiles, or how reptiles could change to mammals. Rather, the revelation was that the idea of creation held by the modern mind was suddenly becoming untenable. The crucible for Darwin was not an abundance of positive evidence for evolution but rather negative evidence against creation. Evolutionist Ernst Mayr has pointed out that Darwin's conversion from creationist to materialist was due to three key scientific findings and later reinforced by several additional findings. These scientific findings were all findings against creation.[17] In other words, the key evidence that swayed Darwin was not direct evidence for evolution but evidence against creation that indirectly argued for evolution.

As Mayr further points out, the doctrine of fixity of species was a key barrier to overcome if the concept of evolution was to flourish:

> Darwin called his great work *On the Origin of Species*, for he was fully conscious of the fact that the change from one species into another was the most fundamental problem of evolution. The fixed, essentialistic species was the fortress to be stormed and destroyed; once this had been accomplished, evolutionary thinking rushed through the breach like a flood through a break in a dike.[18]

The pre-Darwinian metaphysic was that species were fixed and essentialistic. Evidence for small-scale change argued against the old view and in so doing became important evidence *for* evolution.

God Is Not a Micromanager

Darwinism, which called for all species to result from natural speciation events, was the very antithesis of fixity of species. Obviously, evidence against fixity was critical for Darwin and remains so for his modern disciples. It matters little that small-scale change offers limited positive scientific support for evolution. This evidence is, and always has been, critical in evolution's triumph over the doctrine of creation, so it is no surprise that Mayr, Asimov, and others trumpet examples of small-scale change as proof *for* evolution.

An important piece of evidence against fixity of species was the overflowing abundance of species that defied Linnaeus's neat, clean hierarchical classification system. Darwin considered this problem in chapter

2 of *Origin*. He cited examples of how this abundance leads to disputes among botanists trying to decide whether newly discovered forms represent new species or are merely varieties of a species.

The great Swiss botanist Augustin de Candolle (1778–1841) developed his own "natural" classification system which, over against the Linnean "artificial" system, was not wed to the fixity of species concept but allowed for intermediate states. De Candolle's massive study of oaks led him to conclude that most species cannot be clearly defined. According to de Candolle, we can fall prey to the misconception that barriers between allied species are well defined only when we have not fully explored the genus in question. But as we come to know the various species in the genus, we increasingly find intermediate forms that seem to fill in the gaps.[19] The Linnean hierarchical system, it seemed, arose not from a thorough study of all nature but from an initial cursory peek.

Darwin posed a parable of a young naturalist who has difficulty classifying the wide variety of forms he discovers in the field, only to be even further perplexed when allied forms from around the world are brought home for study. Darwin concluded that no clear line of demarcation was evident between species and subspecies. As for de Candolle's three-hundred-odd species of oak, the majority were provisional, perhaps not representing true species at all. Could a divine hand be behind all this confusion? Apparently not; as Darwin announced, de Candolle "no longer believes that species are immutable creations, but concludes that the derivative theory is the most natural one."[20]

It seemed improbable that God would create such a menagerie of oaks, and Darwin applied the same logic to the beaks of birds and even our own beaks. His friend Charles Lyell doubted that human beings could have evolved, so Darwin asked if Lyell could believe that "the shape of my nose was designed." If so, Darwin joked, "I have nothing more to say."[21] Darwin's quip bespeaks how intertwined his theory of evolution had become with metaphysics. Not only did small-scale change argue against fixity of species, but God could never be expected to micromanage creation anyway.

Evidence for Evolution Incorporates Religious Ideas

More recently evolutionist David Merrell has written, "The creation of new species is, in itself, an insurmountable argument against a static-species concept."[22] Merrell uses the phrase "static-species concept" to refer to the fixity of species, and he is, of course, quite correct. The appearance of new species does falsify the doctrine of fixity of species. But for Merrell this is no mere historical note or side observation; rather, it is taken as strong evidence for evolution.

Similarly, Jonathan Weiner sees small-scale change such as pesticide resistance as evidence against divine creation. He approvingly quotes a biologist who rhetorically asked: "How can you be a creationist farmer any more?"[23] Resistance to antibiotics and pesticides may not prove evolution directly, but don't they prove that species are not fixed?

The problem is that this evidence relies on a particular religious notion. When evolutionists use evidence against fixity of species to lend credence to evolution, they incorporate a particular metaphysical notion into a scientific theory: Evolution is supported by the premise that God must make species absolutely fixed—beaks must not get longer and coloration must not change. And since beaks do get longer and coloration does change, we know that God must not have created them.

In his evolution textbook Ridley compares evolutionary theories with the doctrine of fixity of species. Making an objective comparison, Ridley even uses graphs to model fixity of species and illustrate how it compares to evolution. The comparison is a fair one, but too much importance is attached to the result. For Ridley, the conclusion that fixity of species is false is not a historical note but is used as important evidence for evolution—his discussion of it appears in the chapter titled "The Evidence for Evolution."[24]

Evolutionist Tim Berra tells of a rabbit that developed resistance to a deadly virus while the virus simultaneously evolved nonlethal strains because the deadly versions could not spread as easily. Berra concludes not that such a development directly supports evolution but that "it is not explainable by any other concept."[25] As Mayr writes, it is the exposure of weaknesses in opposing theories that has helped evolution the most:

> The greatest triumph of Darwinism is that the theory of natural selection, for 80 years after 1859 a minority opinion, is now the prevailing explanation of evolutionary change. It must be admitted, however, that it has achieved this position less by the amount of irrefutable proofs it has been able to present than by the default of all the opposing theories.[26]

Of course evolution properly advances by the failure of other *scientific* alternatives, and some of the alternatives have indeed been scientific. But the compelling supports for evolution have come from arguments against the doctrine of creation. Unfortunately, evolutionists' rebuttals to creation, though cloaked in scientific terms, are metaphysical because they hinge on one's doctrine of God and creation.

4

The Fossil Record

When the layperson thinks of evolution, it is usually fossils that come to mind. You can hold them in your hand and scrutinize them with the naked eye. The fossil record tells us that life was once different. Today's species are not always represented in the ancient geological strata; a variety of otherwise unknown forms takes their place.

Are these fossilized species the forerunners of the modern species? Is the primitive fish, for example, a distant ancestor of today's giraffe? Evolutionists think so, and they argue that the fossil record is part of the evidence for these ancient relationships. This chapter reviews a representative set of this evidence, the problems with this evidence, and finally the metaphysical arguments to which evolutionists often resort.

The Evidence

Just two years after Darwin published *Origin*, the first of a handful of remarkable reptile-bird intermediate specimens was found in a Bavarian slate quarry. The specimen was named *Archaeopteryx* and was estimated to date back to the Jurassic, about 150 million years ago. Its reptilian features included toothed jaws, clawed fingers, abdominal ribs, and an elongated bony tail. Its avian features included feathers, a furcula (wishbone), and a bird's pelvis.

Large–Scale Change

Evolutionists believe that reptiles evolved into birds and that as this transition occurred nature produced creatures that, like *Archaeopteryx*, looked a bit like a reptile and a bit like a bird. In other words, the theory of evolution predicts transitional fossil species that reveal the pattern of change as one species evolved into another. *Archaeopteryx* has been hailed as an example of fossil evidence in support of evolution. Ironically, paleontologists now question whether *Archaeopteryx* was actually an ancestor of today's birds. Nonetheless, though it may not be the ancestor of modern birds, paleontologists believe it demonstrates the transition. *Archaeopteryx*, or something like it, they say, descended from reptiles and led to birds.

Also following shortly after *Origin* was a reconstruction of the history of the horse. It was the first such lineage to be developed and remains probably the most famous. In 1870 Thomas Huxley announced to the Geological Society in London that he had found evidence for Darwin's evolution that would stand up to rigorous criticism: the European horse lineage. A few years later Huxley met with Othniel C. Marsh, paleontologist at Yale University. Marsh and his rival Edward Drinker Cope had reconstructed what they thought was an unbroken sequence based on North American fossils. The North American fossils were arranged by Henry Fairfield Osborne, director of the American Museum of Natural History. The main lines of evidence were increasing overall size (a concept to be canonized as Cope's Law), decreasing number of toes, and enhancement of grinding teeth. Marsh convinced Huxley that Huxley's European genealogy was mistakenly based on a series of migrations and did not represent a true evolutionary lineage. Marsh believed that the line of descent was gradual, linear, and direct and that specimens of all the important transitions were in hand, from the dog-size *Eohippus* (dawn horse) of the Eocene to the modern horse. Marsh's reconstruction of equestrian evolution prevailed for much of the twentieth century.

Paleontologists now believe, however, that horse fossils reveal a much more complex history. Instead of one species gradually blending into the next improved version, the fossils reveal a spectrum of discrete varieties. New species coexist with older ones, and species seem to emerge rapidly.

Despite the varying versions of horse history, evolutionists point to the horse reconstruction as an impressive example of transitional forms. It is based on thousands of fossils gathered from five continents, spanning an estimated fifty-five million years.

Along with the horse genealogy, evolutionists claim the whale genealogy as important evidence. Whales are thought to derive from the early ungulates—herbivorous hoofed mammals. A candidate ancestor from the early Eocene is the fossil specimen *Hapalodectes*, a small land animal that could swim by paddling its feet. But *Hapalodectes*, estimated to be fifty-five million years old, may well be derived from the whale ancestor rather than standing directly in the whale lineage.

The oldest fossil that is considered directly ancestral to the whale is *Pakicetus*, from the early to mid-Eocene, fifty-two million years ago. It was not a deep diver, did not have the whale's famous blowhole, and may have been amphibious. Dramatic improvements to the whale sequence reconstruction came with two fossil specimens discovered in the early 1990s. *Ambulocetus natans*, thought to be from the early to mid-Eocene, fifty million years ago, was something like the sea lion. The nearly complete fossil looks like a walking and swimming whale, although the pelvis was not found so it is not known how the legs attached to the rest of the skeleton.[1]

Rodhocetus, assigned to the mid-Eocene, forty-six million years ago, had smaller hind legs, although it could likely still move awkwardly on land. After these come *Basilosaurus isis*, thought to be from the late Eocene, forty-two million years ago, with yet smaller legs. It is thought to be related to modern whales but not directly ancestral. The sequence is completed with a series of "archeocete whales," increasing in size and eventually losing the hind legs.

As a final example, the transition from synapsid reptiles to mammals is the best-documented anatomical transition between vertebrate classes. This is a major transition that includes an expanded brain, limbs moved underneath the abdomen, and a remodeled jaw. The jaw transition is a key anatomical change that evolutionists consider particularly well documented. The reptilian jaw contains several bones, whereas the mammalian jaw consists of a single bone. Paleontologists have reconstructed what for Stephen Jay Gould is a "lovely sequence of intermediates,"[2] where the reptilian jaw bones move aft and become the mammalian middle ear. *Cynognathus*, estimated to date from the early Triassic, at least 240 million years ago, is a key specimen in the sequence. It sports two jaw joints corresponding to the old reptilian articulation and the new mammalian connection.

Small-Scale Change

We have reviewed four major lineages that paleontologists believe reveal large-scale evolutionary change at work. There are also plenty of more detailed lineages that paleontologists believe evidence small-scale change. Diatoms are single-celled photosynthetic organisms that float on water. A rather complete fossil record is available revealing two different-sized diatoms that appear to derive from a single intermediate. The fossils span an estimated 1.7 million years, and the size is inferred through measurement of the height of the hyaline, a glasslike area of the cell wall. The two sizes are about two and five micrometers in height for most of the time window, but they converge to a similar value early in the window.[3]

An exceptionally complete sedimentary record at Lake Turkana in Kenya has yielded an abundance of snail specimens. More than three thousand specimens, from the Pliocene and Pleistocene, one to five million years ago, were analyzed. Twenty-four different characters were measured from thirteen lineages. The results revealed a stepwise sort of evolution, where species remained unchanged for long intervals and then abruptly changed at points when the lake level is believed to have changed.[4]

Another exhaustive study involved more than three thousand specimens of trilobites from eight different generic lineages. The fossils came from seven stratigraphic sections estimated to be about five hundred million years old and to span about three million years. The numbers of pygidial ribs were recorded, and each lineage showed slow, gradual change over the time window.[5]

Problems with the Evidence

These examples are evidence for evolution. In fact, for many evolutionists such examples are tantamount to proof of the theory. This is the hard evidence that cannot be denied, they claim. The National Academy of Sciences concludes that the fossil record "provides consistent evidence of systematic change through time—of descent with modification."[6] But "systematic change through time" is a scientific description, whereas "descent with modification" is a particular explanation for the observation. The former does not prove the latter. This section examines the problems with the evidence and cautions against such an optimistic conclusion.

Large Changes Occur Rapidly

One problem with the fossil evidence is its abrupt character. If we are to believe that evolution occurred, then according to the fossil record large evolutionary change probably happened in relatively short periods, with little or no change in between. Consider, for example, the origin of the first organic cells. The early Earth was inundated by meteors that wreaked havoc on a global scale. Recent studies have shown that this process took the better part of a billion years, pushing right up to the estimated date of the earliest fossils showing signs of primitive life. The difference looks to be less than ten million years. In other words, evolution may have had no more than ten million years to produce the first organic cells. This is a remarkably short time frame, especially considering how far back in time we are peering. It is also a short time for the evolution of the basic unit of life—the cell—which even in its most primitive form is highly complicated.

But this is only the first of many "big bangs" of biology. Paleontologists estimate that over the past 600 million years the major groups in the fossil record made abrupt appearances. Despite the transitional sequences given as examples above, the fossil record generally seems to suggest long periods of boredom with short fire drills interspersed.

There is, for example, the rapid appearance of many plants. "Nothing is more extraordinary in the history of the Vegetable Kingdom," wrote Darwin in a letter to a friend, "than the apparently very sudden or abrupt development of the higher plants."[7]

As one recent paleontology text put it, "The observed fossil pattern is invariably not compatible with a gradualistic evolutionary process." There is a problem either with the fossil record or with the idea that evolution is gradual. To make the data compatible with the theory, "undiscovered fossil forms can be proposed" or "unknown mechanisms of evolution can be proposed." But neither of these ad hoc hypotheses is known to be true or untrue.[8]

Such ad hoc hypotheses are often used by evolutionists to try to explain the "Cambrian Explosion," the most spectacular of biology's big bangs. Estimated to have taken place almost 600 million years ago over a period no greater than five million years, it initiated virtually all the major designs of multicellular life with barely a trace of evolutionary history. In a geological moment, the fossil species went from small worm-like creatures and the like to a tremendous diversity of complex life forms, including virtually all of today's modern designs.

Evolutionists have had little success explaining this one-time event. Thomas H. Huxley likened it to a barrel that is filled rapidly with apples, after which it takes longer to fill the remaining spaces with pebbles, sand, and finally water.[9] Today's explanations are more technical-sounding but no less reliant on speculation as opposed to direct description. Steven M. Stanley compared the Cambrian Explosion to the introduction of bacteria croppers that prey on dominant species that previously had suppressed diversity. J. J. Sepkoski compared it to rapid growth of bacterial populations in a virgin petri dish. Were the Precambrian oceans a virgin ecosystem with the raw materials of oxygen and food supplied by ancient bacteria?[10] Geneticist Steve Jones wonders if the Cambrian Explosion reflects some crucial change in DNA—life's genetic material. "Might a great burst of genetic creativity," asks Jones, "have driven a Cambrian Genesis and given birth to the modern world?"[11] Of course, any of these speculations is possible, but evolution gives us very little explicit explanation of precisely how and why life took this dramatic turn.

Evolution did not predict, nor can it provide a detailed explanation for, abruptness in the fossil record. But evolutionists are not alarmed, for the Cambrian Explosion does not refute evolution. They point out that observed rates of small-scale change are sufficient to account for the abrupt changes observed in the fossil record. None of biology's big bangs, they say, were so rapid that evolution could not possibly have produced the observed change.

And how do evolutionists measure these rates of change? They measure rates of small-scale changes within species. For example, traits in guppies, such as growth patterns, were found to change when the guppies were placed in a new environment. The guppies, of course, were still guppies, but evolutionists argue that the *rate of change* observed is theoretically sufficient to account for any of the abrupt changes seen in the fossil record.[12] We could argue, against the evolutionists, that there is no justification for assuming that such small-scale changes fall into the same category as large-scale changes. But it is important here to understand the thrust of the evolutionists' argument. They are not showing that evolution is compelling or even likely; they are merely saying that evolution is not proved false by abruptness in the fossil record.

It certainly is true that one cannot use biology's big bangs to absolutely disprove evolution, but this simply points out how adaptable evolution is to whatever evidence comes along. One might think that evolution requires evidence of slow, gradual change, but in fact evolution can also accommodate abruptness in the fossil record. Why should we accept a theory that does not provide compelling explanations or bold predictions but rather molds itself to whatever evidence comes along?

Rather than being falsified by abruptness, evolution simply adopts it. We are told that big bangs such as the Cambrian Explosion do not call evolution into question; they define evolution. They help answer the question *how* evolution occurred, not *whether* it occurred. Darwin was concerned about the abruptness of the fossil record, but according to Eldredge, he was simply confusing "the basic notion that evolution has occurred with the various possible ways in which evolution might conceivably happen."[13] For Jones, in fact, the Cambrian Explosion is a failure not of Darwin's theory but of the fossil record. Yes, for some reason shells appeared all of a sudden, but they must have evolved from soft creatures that leave no mark on the geological record.[14]

The fact that the Cambrian Explosion does not refute evolution does not mean that the abruptness problem is resolved. There are all sorts of unlikely theories that otherwise cannot be falsified. What science needs are likely explanations for its observations. And the array of vague explanations about how evolution could have produced big bangs such as the Cambrian Explosion does little to help. Their speculative nature reveals what little hard evidence there is that evolution is the right explanation and, in spite of what evolutionists maintain, how big a problem the Cambrian Explosion is for evolution.

Too Much Complexity

The apparent abruptness with which species appear in the fossil record is especially puzzling in light of their complexity. One might think that in the evolution of a new species, simple forms would appear first and then complexity would develop, building on the early successes. Of course, this is the overall trend in the fossil record, with simple bacteria in the beginning, almost four billion years ago, and *Homo sapiens* at the tail end. But fossil species are complex when they first appear and often tend to remain unchanged from that point on. In other words, we see a progression of highly developed, complex forms more than a building up of complexity.

Consider the eye. The dozens of different types of eyes found in nature did not evolve from a common, primitive version. Evolutionists believe they must have evolved independently because an evolutionary sequence cannot be drawn between them. A remarkable example comes from the ancient and humble trilobite from half a billion years ago. Its eyes were perhaps the most complex ever produced by nature.[15] One expert calls them "an all-time feat of function optimization."[16]

Of course, the fossil record cannot usually tell us about soft body parts or the behavior of its specimens. For these we look to extant species, and here a brief excursus is appropriate to review a few examples so complex that they challenge evolution as an explanation for their existence. Certain types of bats map out objects as small as mosquitoes by sensing the echoes of their own squeaks—a system known as echolocation. To begin with, the bat emits a high-pitch squeak, well beyond the range of human hearing, up to two thousand times per second. Next it determines both range and direction to the tiny mosquito by sensing the echo while filtering out echoes from the squeaks of nearby bats. Or consider fish that use underwater electric fields either passively or actively to sense objects around them, including other fish. The details of such systems would fill books. Anyone familiar with today's sonar or radar systems knows the immense complexity inherent in such systems: the problems of sensing the echo in the presence of the transmitted signal, which can be billions of times stronger, of filtering out spurious signals such as echoes of older transmissions, of combining the echo information with knowledge of one's own motion, and so forth. Yet the bat's detection abilities are superior to those of the best electronic sonar equipment.

Or consider the baffling and complex behaviors found in many organisms. Certain species of the *Hydra*, a small underwater creature, develop nematocysts—stinging cells that eject a tiny poisoned hair. Meanwhile, a planarian worm known as the *Microstomum* consumes *Hydra* but passes the nematocysts through its digestive system and positions them on its surface. The *Hydra* meal serves to arm the *Microstomum*, and when fully equipped the *Microstomum* omits the *Hydra* from its diet, resuming feeding on it after discharging its ill-gotten arsenal.[17]

For evolution to have formed this system, certain *Microstomum* must have happened to have selectively digested the *Hydra*, leaving the nematocysts untouched. Then they happened to have vectored the nematocysts to the surface and positioned them there. Then certain *Microstomum* happened to have a feedback loop installed to regulate their diet.

Or consider a sheep parasite known as the brainworm. Robert Wesson explains its fascinating life cycle:

> The brain worm that reproduces in sheep uses ants to get back into a sheep. The worms get into ants by infecting snails that eat sheep feces. The snails expel tiny worm larvae in a mucus that ants enjoy, and some dozens of worms take up residence in an ant. But this would do them no good if the ant behaved normally; too few ants would be eaten by sheep. Consequently, while most of the worms make themselves at home in an ant's abdomen, one finds its way to the ant's brain and causes the ant to climb up a grass stem and wait to be eaten by a sheep. Ironically, the worm that programs the ant is cheated of happiness in the sheep's intestine; it becomes encysted and dies.

The whole procedure seems unnecessary. Why do the worm eggs defecated by the sheep not simply hatch and climb up the grass stem to await being eaten by a sheep instead of making the hazardous trip through snail and ant? How could they become adapted to being carried by the ant unless the ant were already programmed to make itself available to be eaten by a sheep?[18]

Then there is the bombardier beetle that secretes an irritating chemical solution to fend off enemies. When danger lurks, the beetle generates the irritant by mixing hydrogen peroxide, hydroquinones, and appropriate enzymes. The beetle has its own miniature chemical warfare system, which defies Darwin's notion of unguided evolution.

The list, of course, goes on and on. The decoy fish has a detachable dorsal fin that mimics a smaller fish, complete with a dark spot resembling an eye and notch resembling a mouth. The decoy fish becomes motionless except for the decoy, which moves from side to side, causing the "mouth" to open and close. And there is the owl with ears tuned to different frequencies, to better track its prey, and the rattlesnake with heat-sensitive (infrared) sensors to image its prey at night.

At the molecular level there also are mechanisms that seem to defy any sort of gradual or stepwise evolution. Examples of these are legion; Michael J. Behe picks out five that exemplify what he calls *irreducible complexity*. A system is irreducibly complex if it relies on all of its parts in order to function. Such a system could not have evolved gradually by adding its component parts one at a time, because the system doesn't work without all the parts working together at once.[19]

To be sure, evolutionists have worked hard on the problem of how complex systems could arise from the blind, undirected process of evolution. In fact, according to them the problem is all but solved. They claim that they have successfully shown how many such systems could have evolved. Have they? As we shall see in the next two sections, the answer is yes only if we settle for rather vague explanations.

Evolution Relies on Speculative Explanations

Complexity is a problem for evolution. What is the source of superior forms that require complex interaction of disparate mechanisms and behaviors? What good, for example, is half a wing? One solution is the *change of function principle*, which finds all sorts of possible uses for a partial wing. It could be used for warming the creature, physical protection, fighting, sexual display, or gliding, just to name a few. Thus the ancestral bird is *preadapted* for flight by virtue of its possession of feathers.

The problem is not finding beneficial applications for half-baked mechanisms; rather, it is deciding which application was actually used. Regarding the bombardier beetle, Darwin disciple Richard Dawkins explains: "As for the evolutionary precursors of the system, both hydrogen peroxide and various kinds of quinones are used for other purposes in body chemistry. The bombardier beetle's ancestors simply pressed into different service chemicals that already happened to be around. That's often how evolution works."[20]

Dawkins makes it sound so easy, but such an explanation is speculative. Others have put forth their own explanations for the bombardier beetle,[21] but they too ultimately fall back on Dawkins's heroic assumption that natural selection "simply pressed into different service chemicals that already happened to be around."

Evolution Settles for Plausible Explanations

Although the intricacies of the bombardier beetle and many of our other examples were yet to be discovered in the nineteenth century, the problem of complexity was no less felt by Darwin and his fellow naturalists. He wrote in *Origin*:

> To suppose that the eye, with all its inimitable contrivances for adjusting the focus to different distances, for admitting different amounts of light, and for the correction of spherical and chromatic aberration, could have been formed by natural selection, seems, I freely confess, absurd in the highest possible degree.[22]

But Darwin then argued that we must not be misled by our intuitions, which are highly susceptible to misconception. For given natural selection operating on inheritable variations, some of which are useful, then if a sequence of numerous small changes from a simple and imperfect eye to one complex and perfect can be shown to exist, and if the eye is somehow useful at each step, the difficulty is resolved.[23]

Here Darwin lays out the form of the preadaptation argument. What is needed, according to Darwin, is a *conceivable* sequence. If evolutionists, by thought experiment, can conjure up any sequence that shows a potential usefulness at each stage, then the problem is solved. We need not pursue what *likely* happened; what *could have* happened will do. Whether one finds Darwin's solution to the problem of complexity satisfying or not, he pointed out that it is a valid argument:

> Although the belief that an organ so perfect as the eye could have been formed by natural selection, is enough to stagger any one; yet in the case

of any organ, if we know of a long series of gradations in complexity, each good for its possessor, then, under changing conditions of life, there is no logical impossibility in the acquirement of any conceivable degree of perfection through natural selection.[24]

The only required premise is that "we know of a long series of gradations in complexity, each good for its possessor." Of course, what Darwin intends here is that we simply must be able to envision such a sequence. But one can always, by thought experiment, conjure up a set of potentially useful intermediates. Thus, while it is true that there is no "logical impossibility" to Darwin's solution, we must also say that it is not falsifiable. How could a would-be critic show that *no such sequence* exists?

As Darwin put it: "If it could be demonstrated that any complex organ existed, *which could not possibly have been formed* by numerous, successive, slight modifications, my theory would absolutely break down. But I can find out no such case."[25] But this was hardly a concession. Darwin may sound generous here, allowing that his theory would "absolutely break down," but his requirement for such a failure is no less than impossible. For no one can show that an organ "could not possibly" have been formed in such a way. So in short order Darwin reduced what seemed to be a dilemma for his theory into a logical truism. Evolution was protected from criticism, and all that was needed to explain complexity was a clever thought experiment.

Darwin so lowered the requirements that anyone with a pen and a vivid imagination can now claim to have solved the problem of complexity. It is now common to see in the evolution literature vague explanations, relying on such dubious mechanisms as "chance" or "opportunism," put forth as though they were solutions to the problem of complexity.[26] These solutions simply do not support the often-made claims that complexity is not a problem for evolution. Along with short time windows and abruptness, the problem of complexity remains unresolved.

Are Transitional Forms Really Transitional?

But what about the positive evidence for evolution we saw at the beginning of this chapter? How important can the problems of small time windows, abruptness, and complexity be with such strong evidence in hand? Or could the positive evidence be weaker than is sometimes claimed?

The fossil patterns raise questions about the conclusion that they are part of an evolutionary process. *Archaeopteryx*, for example, appears to be related to both reptiles and birds, but we have little in the way of candidates for its reptilian predecessor or its avian descendants. In the case of the whales, we have a set of specimens that cannot be aligned in a single sequence, and the supposed evolutionary relationships remain unresolved. The specimens are points in a bush that remains to be filled in rather than a lineage nearing completion. And in any case convergent evolution is ultimately required as an explanatory device. Are these specimens really transitional forms leading to the modern whale, or are they simply different species that have been so arranged in our imagination?

Furthermore, the evolution story leaves us to wonder in awe at the supposed transition that has the whale's ancestors losing their hind limbs, grinding teeth, and pelvises and developing a host of new features in very little time with great efficiency. The new features include the fluke tail with its unique vertical propelling motion, the huge filter-feeding jaw, and the ability to give live birth and raise young in the marine environment. Yet despite this tall order, evolution proposes that in a relatively short period the ancestral land animal completed the journey back to the ocean and acquired superior skills in its new marine environment. The latest entry to the community could swim, dive, and feed as well as or better than most fish and sharks.[27] All sorts of evolutionary scenarios can explain why the whale acquired such advanced skills, but they are speculative. The whale's aquatic prowess does not refute evolution, but it raises the question of how we can be so sure about the evolutionary change that is supposed to have created the whale.

The horse lineage, although full of horselike fossil specimens, nevertheless leaves paleontologists wondering whether the history can be accounted for by neo-Darwinian gradualism—the extrapolation of small-scale change to large-scale change—or whether we need an additional large-scale change mechanism to account for steps in the fossil record.[28] For each of the various species seems to persist unchanged for millions of years only to have a new species appear alongside it, coexisting until the first finally becomes extinct. Eldredge writes that horses "got larger, lost the side toes on their feet, and evolved progressively larger and more complicated teeth (for grazing)," yet "there is little evidence of gradual progressive change of the sort we would expect from the operation of pure natural selection. What we see, again, is persistence of species once they appear—and persistence in a virtually unchanged condition."[29]

There does not appear to be a distinctive sequence that could be used to imagine the evolution of the horselike species as first was thought. An untold number of textbooks and museum exhibits touted the horse sequence as clear evidence for evolution, but now the story isn't so obvious. This has prompted evolutionists to change their metaphor from a tree to a bush. But why should we take these fossil specimens as evidence for Darwin's descent with modification at all? The bush metaphor reminds us of the fact that the fossil record gives a historical record of species but does not tell us *how* the species got there.

The reptile-mammal transitional sequence also relies on the bush metaphor. Laurie R. Godfrey says, "We *have* found clear evidence of transitional links between major vertebrate groups." There are so many links, in fact, that "in many cases, the problem is not a lack of intermediates but the existence of so many closely related intermediate forms that it is notoriously difficult to decipher true ancestral-descendant relationships."[30] Douglas Futuyma echos this sentiment: "The gradual transition from therapsid reptiles to mammals is so abundantly documented by scores of species in every stage of transition that it is impossible to tell which therapsid species were the actual ancestors of modern mammals."[31]

If it is "notoriously difficult to decipher true ancestral-descendant relationships," then how can evolutionists be so sure there is one? Certainly we can select our favorite sequence, but the fossils cannot tell us which is the correct sequence, or even whether there is a correct sequence at all.

If we pick and choose from the abundant pool of available fossils to synthesize a sequence, we may be creating our own reality instead of reconstructing the true history of life. In fact, with evolution we must believe that across the reptile-mammal transition organisms evolved so rapidly that they appear fully formed and diverse in the fossil record;[32] that there are large gaps between the reptiles and mammals;[33] and that convergent evolution must have occurred many times.[34]

Small–Scale Surprises

We saw that the fossil record sometimes records detailed changes in species, but in these cases there is no evidence that the observed changes could ever add up to the large-scale change evolution requires. In the examples above, diatoms, snails, and trilobites underwent morphological change, but they were still diatoms, snails, and trilobites. Regarding the trilobites, Gould and Eldredge point out that such small changes do not help characterize evolutionary change except minor changes in the varieties within a species. Indeed, when new forms above the species level appear in the trilobite record, they do so rather abruptly.[35]

In one trilobite study Eldredge found a bizarre menagerie of unexplainable forms, including spines on the front, sides, and middle of the head and around the margins of the tail. The trilobites also failed to show signs of an evolutionary trail. Eldredge found most of the different kinds present in the earliest-known fossil beds, including some of the unique and advanced ones. And there was no evidence for gradual change in the older species that would have allowed Eldredge to predict the anatomical features of its descendants. As Eldredge explains:

> Standard evolutionary theory focuses on anatomical change through time by picturing natural selection as the agent that preserves the best of the designs available for coping with the environment. This generation by generation process, working on small amounts of variation, is thought to change, slowly but inexorably, the genetic and anatomical makeup of a population.
>
> If this theory were correct, then I should have found evidence of this smooth progression in the vast numbers of Bolivian fossil trilobites I studied. I should have found species gradually changing through time, with smoothly intermediate forms connecting descendant species to their ancestors.
>
> Instead I found most of the various kinds, including some unique and advanced ones, present in the earliest known fossil beds. Species persisted for long periods of time without change. When they were replaced by similar, related (presumably descendant) species, I saw no gradual change in the older species that would have allowed me to predict the anatomical features of its younger relative.
>
> The story of anatomical change through time that I read in the Devonian trilobites of Gondwana is similar to the picture emerging elsewhere in the fossil record: long periods of little or no change, followed by the appearance of anatomically modified descendants, usually with no smoothly intergradational forms in evidence.
>
> If the evidence conflicts with theoretical predictions, something must be wrong with the theory. But for years the apparent lack of progressive change within fossil species has been ignored or else the evidence—*not* the theory—has been attacked. Attempts to salvage evolutionary theory have been made by claiming that the pattern of stepwise change usually seen in fossils reflects a poor, spotty fossil record. Were the record sufficiently complete, goes the claim, we would see the expected pattern of gradational change. But there are too many examples of this pattern of stepwise change to ignore it any longer. It is time to reexamine evolutionary theory itself.
>
> There is probably little wrong with the notion of natural selection as a means of modifying the genetics of a species through time, although it is difficult to put it to the test. But the predicted gradual accumulation of change within species is seldom (if ever) encountered in our practical experience with the fossil record.[36]

Eldredge found that trilobite fossils did not fit "standard evolutionary theory." For him the theory of evolution was not wrong; it was simply in need of some adjustment. One such adjustment is punctuated equilibrium, proposed by Eldredge and Gould in 1972. The punctuated equilibrium framework recognizes that the fossil record reveals long periods of no change and short periods of large change. This pattern in the fossil record was well known in Darwin's time. His friend and supporter Thomas Huxley was concerned that Darwin had not incorporated it into his theory. "You have loaded yourself," warned Huxley, "with an unnecessary difficulty in adopting *Natura non facit saltum* [nature does not make leaps] so unreservedly."[37]

But Darwin had a more synoptic view of his new paradigm. He had used negative theology to argue that there was no divine hand in nature. Because of nature's quandaries, Darwin argued, we should look not to the Creator but to natural laws. And whereas the Creator might adjust his creation in an instant, natural laws operate continuously over time. It is true that natural laws could occasionally add up to abrupt changes, such as in earthquakes. But to admit that large-scale jumps were of general importance in creating the species would open the door to purpose and design. For why should such large-scale jumps be ascribed only to natural laws?[38]

Darwin had crafted his argument carefully and was more attuned to the various skeptics than was Huxley. For instance, geology's catastrophists had viewed sharp changes in natural history as evidence of the Creator's forming the world. The uniformitarianists had overcome the catastrophists and in many ways had assisted Darwin (see chapter 7). He could hardly give up that hard-fought ground without paying a heavy price. Yes, the fossil record suggested that nature takes jumps, but it was safer for Darwin to question the data than to admit it into his theory:

> The geological record is extremely imperfect and this fact will to a large extent explain why we do not find interminable varieties, connecting together all the extinct and existing forms of life by the finest graduated steps. He who rejects these views on the nature of the geological record, will rightly reject my whole theory.[39]

In order for evolution to succeed, Darwin would need to steer clear of the supernatural, or anything that could be interpreted as supernatural, and argue for a strictly naturalistic origin of species. Darwin could hardly argue that "God wouldn't have done it that way" and then propose a theory that allowed for a creationist interpretation.

In its first century, evolution maintained Darwin's hope that the fossil record was incomplete, and most evolutionists carefully stepped around the problem of stasis and abruptness in the fossil record. But after a hundred years it was safe to question the dogma. In the spirit of Huxley, punctuated equilibrium acknowledges the fossil record's pattern of stasis and abruptness and tries to incorporate it into evolution. The difference is that after more than a century, evolution has now secured its place in science. Darwin's concern that God be distanced from nature is now taken for granted. With God out of the picture, evolutionists can now approach the problem of nature's discontinuities.

The result is that punctuated equilibrium now embraces the fossil record pattern. Favorable variations are envisioned to spread quickly as speciation is hypothesized to occur in isolated pockets. This fast-paced process would leave few fossil remains, and the fossils that are preserved would come from large central populations where little or no change takes place. "Thus, the fossil record," Gould concludes 120 years after Darwin, "is a faithful rendering of what evolutionary theory predicts."[40] Darwin's theory has now been carefully tailored so as to produce exactly what we find in the fossil record.

One might wonder what evidence gave Gould and Eldredge this new insight into an old problem. For the most part it was the fossil record itself. The fossil record reveals a nongradual pattern, so evolution must proceed this way. The evidence has continued to accumulate. Paleontologist Alan Cheetham rejected punctuated equilibrium when it was first proposed; he considered himself a gradualist. But his exhaustive study of the remains of Bryozoa (coral-like animals), dated over the past fifteen million years, convinced him otherwise. Over and over he saw that change was sudden and individual species persisted virtually unchanged for two to six million years.[41]

With punctuated equilibrium added to gradualism, Darwin's theory of evolution has been expanded to the point where it can explain practically anything. Slow change, fast change, no change, and even reverse change can all be given the proper label. But what this really shows is how adaptable evolution is to whatever evidence comes along.

The various problems covered here do not completely refute evolution, but then again there are all sorts of unlikely ideas that cannot be completely refuted. What these problems do show is that the fossil data are not strong evidence for evolution.

Metaphysical Arguments

Despite the unresolved issues in the fossil record, evolutionists routinely claim that fossil evidence proves evolution. For example, Berra claims that "Fossils provide hard evidence that evolution has occurred,"[42] and the fact that "even one transitional fossil is found is a sufficient demonstration of evolution and a resounding falsification of creationism."[43] Likewise, in 1951 George G. Simpson wrote that "there really is no point nowadays in continuing to collect and to study fossils simply to determine whether or not evolution is a fact. The question has been decisively answered in the affirmative."[44] More recently, Kenneth R. Miller concludes that the fossil data indicate that "descent with modification, which most of us prefer to call evolution, really happened."[45]

How is it that evolutionists have become so confident? How is it that the fossil evidence is taken as proof for the theory of evolution? From the writings of evolutionists themselves the answer is obvious. Fossils, like the small-scale change and comparative anatomy evidences, seem to disprove the theory of divine creation. If supernatural explanations are ruled out, then by the process of elimination only naturalistic explanations remain.

God Would Not Have Made This World

The fossil record may not provide unequivocal support for evolution, but evolutionists argue that it reveals a world that God would not have created. First, why would God create so many different types of species? The sheer volume and timing of species call for an incredible level of divine activity. To create the billions of species that have been discovered, the Creator must have been constantly generating new "kinds," especially in the last 600 million years. We must believe that, as the geneticist J. B. S. Haldane (1892–1964) remarked, the Creator has an inordinate fondness for beetles to have created over 250,000 different species.[46] Or as the great science writer Martin Gardner bluntly puts it, "Because there are millions of insect species alone, this requires God to perform many millions of miracles. I cannot believe that."[47]

Furthermore, of those innumerable creation acts, the vast majority were to result in extinctions. What could have been the Creator's intentions with these failed attempts? The eighteenth century grappled with this problem as it became increasingly evident that species had indeed failed to survive. Modern thinkers as disparate as John Wesley and Thomas Jefferson had agreed that their good God never would have allowed species to become extinct. Death, said Wesley, "is never permitted to destroy the most inconsiderable species." For Jefferson, no link in the great work of nature could be so weak as to be broken.[48] These ideas translate into yet more evidence for evolutionists' theory. Sir Gavin de Beer, for example, concluded that unless "one is prepared to believe in successive acts of creation and successive catastrophes resulting in their obliteration, there is already a strong presumptive indication that evolution has occurred."[49] Likewise, for Miller, without evolution we are left with a designer who "just can't get it right the first time. Nothing he designs is able to make it over the long term."[50]

In spite of the sorry fate of so many of the species, many must have been created in a hierarchical sequence between the successive creations. This results in ancient fossils that appear strange to us and recent fossils that are more familiar to us—a trend that for Futuyma argues against creation:

> As we pass from the remotest periods of geological time toward the present, the fossils become more and more modern. Certainly some groups such as blue-green algae and horseshoe crabs have persisted since early geological time; but most groups of animals and plants have arisen, flourished, and died out; the most ancient fossils are the strangest to us; and as we approach the present, they get more and more familiar. Jurassic mammals, if they came alive today, would hardly look like mammals to us; by the Cretaceous, we get rather modern-looking opossums; by the Eocene, armadillos; by the Pliocene, modern-looking horses and rhinoceroses. *Regularities of this kind accord with evolution, not creation.*[51]

Futuyma's God would never create the species according to any such order or regularity. Berra also finds it a problem for creation that advanced organisms usually appear later than primitive ones: "This sequential appearance of different groups at different times, the more advanced appearing in general later than the more primitive, is predicted by evolutionary theory. *It cannot be reconciled with creationism.*"[52]

Mark Ridley concludes that fish, amphibians, reptiles, and mammals would not appear in that order in the fossil record if they had been separately created.[53] And for Stephen Jay Gould, God would not have created species in these sorts of sequences and therefore evolution must be true: "What alternative can we suggest to evolution? Would God—for some inscrutable reason, or merely to test our faith—create five species, one after the other . . . , to mimic a continuous trend of evolutionary change?[54]

Who needs scientific evidence when evolution is the only alternative? There is also the problem that fossil species sometimes are similar to living species in the same region. "Why," asks Miller, "should such a unique set of animals be found in exactly the same place as their closest fossil relatives?" Surely God would not create similar species in the same locale. "There could be just one answer," states Miller: "a process of descent with modification linked the present to the past."[55]

But if the fossil record evinces too much order, it also has an *arbitrary* aspect that for some evolutionists does not accord with creation. For the fossil species do not always seem to progress in any particular direction.

"What could have possessed the Creator," asks Futuyma, "to bestow two horns on the African rhinoceroses and only one on the Indian species?"[56] Miller also finds evidence against the Creator in elephant fossils. There are, explains Miller, dozens of elephant or elephantlike fossil species dating back to as much as fifty million years ago. Trends in the design of the trunks and tusks can be found among these species. Using these trends, the species can be compared, classified, and even arranged in an evolutionary tree if one believes in evolution. So we should believe in evolution, according to Miller, for can we possibly believe there is a Creator behind this haphazard arrangement?

> This designer has been busy! And what a stickler for repetitive work! Although no fossil of the Indian elephant has been found that is older than 1 million years, in just the last 4 million years no fewer than nine members of its genus, Elephas, have come and gone. We are asked to believe that each one of these species bears no relation to the next, except in the mind of that unnamed designer whose motivation and imagination are beyond our ability to fathom. Nonetheless, the first time he designed an organism sufficiently similar to the Indian elephant to be placed in the same genus was just 4 million years ago—Elephas ekorensis. Then, in rapid succession, he designed ten (count 'em!) different Elephas species, giving up work only when he had completed Elephas maximus, the sole surviving species.[57]

Miller obviously has specific ideas of what the designer is and is not allowed to do. First, the designer must be sensible to us, going about his work as we see fit. Repetitive work seems unlikely—he certainly wouldn't go about making ten different Elephas species in rapid succession. In fact, if we tally up all the millions of different species ever found, the Creator must have been constantly at work, and this too, for Miller, is hard to believe.[58]

Similar negative theological arguments have also been used for the problem of complexity. True, it may be difficult to imagine how the blind process of evolution could have produced the eye, yet despite its complexity, the eye sometimes seems like a rather haphazard contraption. For example, in the human eye, photoreceptor cells detect incoming rays of light, but the detectors are at the back of the cell. Incoming light must pass through the rest of the cell structure before reaching the detectors. Surely a wise Creator could have come up with a more sensible arrangement. "The feeblest of designers," writes Steve Jones, "could improve it." This and other examples, says Jones, show that complex organs are "not the work of some great composer but of an insensible drudge: an instrument, like all others, built by a tinkerer [i.e., the evolutionary process] rather than by a trained engineer."[59]

A Metaphysical Fact

One could counter Jones with the enormous problem that complexity poses for evolution. And one could counter Ridley, Gould, Miller, and the rest with the various problems with the fossil evidence discussed in this chapter. But for our purposes here, what is important is that evolutionists are using nonscientific arguments for evolution. Their arguments rely on an unspoken premise about the nature of God and how God would go about creating the world.

Certainly it is true that for those who believe in divine creation the fossil record is an insight into the history of creation. And it is natural to wonder why God would do things one way rather than another way. But such speculations are religious, not scientific, for they hinge on one's personal concept of God. Some people find extinctions troubling because they focus on God's benevolence. Others can just as easily interpret extinctions as a result of the futility to which creation has been subjected.[60]

Obviously such speculations should not be included in a scientific debate unless the presuppositions behind them are defined clearly and openly. Unfortunately, this has not been the case with the evolutionists' arguments. They employ assertions about what God would and would not do to prove their point, yet they claim that evolution is a scientific conclusion. The general theme is that the creation hypothesis requires an inordinate amount of providence. Could the Creator possibly have seen fit to create such a menagerie—millions of species of every imaginable design and everything in between? Better to line species up in a sequence and ascribe it to natural law. If God cannot do it, then nature can. And the new view is so convincing because the old view is so untenable. No matter that science is always tentative, metaphysical speculation can claim the truth, and the truth we have. Evolution is a fact for the simple reason that the alternative, modernism's divine creation, is not considered viable.

Evolutionists may forgo natural selection if a replacement mechanism provides a better explanation, but they will not allow for creation. Like Darwin, they are attempting to reconcile the modern view of God with the natural world they confront every day. It is biology's solution to the problem of natural evil.

Now we can understand the sense in which evolution is a fact for evolutionists. They may not be able to tell us how evolution works, but they can tell us how it doesn't work. Evolution by natural means is a fact for the evolutionists simply because creation is impossible. But this whole argument for evolution depends on one's view of God and his creation.

5

One Long Argument

Charles Darwin referred to his book on evolution as one long argument. The three previous chapters of this book have examined the three most important categories of evidence used in that argument, by Darwin and by today's evolutionists. This chapter examines the argument from a different perspective: a historical survey of evolutionists who saw fit to continue with Darwin's long argument. Though the majority of the evolution literature presupposes the fact of evolution, there have from time to time been evolutionists who have attempted to *prove* that evolution is undeniably true—a fact not open to rational debate. This chapter looks at five such works, spanning about a century from 1888 to 1991.

Only works that represent mainstream evolutionary thought were selected for this survey. The survey shows that since Darwin the argument for evolution has not changed very much. The differences between authors are only variations on a central theme. Before the development of modern genetics or the breakthroughs of microbiology, evolutionists were just as convinced as today's evolutionists that evolution is a fact. And all the authors attempt to establish this fact by making arguments for the plausibility of evolution and bolstering these with nonscientific arguments against divine creation.

In chapters 2, 3, and 4 we saw that the best arguments for evolution rely on metaphysical interpretations to make them compelling. This chapter, taking a different perspective, comes to the same conclusion. Looking back through the years, from the late nineteenth century to only a few years ago, one can see that negative theology is woven into the fabric of evolutionary thought. It is not just a sidelight that appears from time to time but is an integral part of evolutionary thinking.

Joseph Le Conte, 1888

Joseph Le Conte was professor of geology and natural history at the University of California at Berkeley and a prominent evolutionist. The first edition of his book *Evolution: Its Nature, Its Evidences, and Its Relation to Religious Thought* was published in 1888. Le Conte was motivated to write a book on evolution because he perceived the evolution literature to be so voluminous, fragmentary, and technical that even very intelligent persons had only a vague idea of the subject.

His book consists of three parts: "What Is Evolution?," "Evidences of the Truth of Evolution," and "The Relation of Evolution to Religious Thought." In the second part Le Conte intends to prove that, in evolution, science has discovered an ultimate truth about the world. He compares evolution with gravity—an analogy that would become popular with later evolutionists—to explain how sure we can be of evolution. He felt that the evidence made evolution so compelling and obvious that those who accept evolution should not be called *evolutionists* any more than those who accept gravity should be called *gravitationalists*. His exposition of the evidence is exhaustive. It is contained in a total of nine chapters—the longest of the works in this survey.

A General Argument for Evolution

Le Conte begins his nine-chapter argument with a general discussion of how the inorganic world evolved by slow, gradual processes. He appeals to the field of geology, which at that time emphasized continuous processes in the earth's natural history. Mountains, continents, rocks, soil, and waterways, he explains, all were formed by a slow "process of evolution."[1] Le Conte also appeals to the field of astronomy, which at that time embraced the view that the universe was infinite. Before the birth of modern astronomy, according to Le Conte, "the intellectual space-horizon of the human mind was bounded substantially by the dimensions of our earth."[2] But astronomy has found space to be infinite and "full of worlds like our own,"[3] all of which originated by a gradual process. Together, Le Conte claims, geology and astronomy have established "the law of universal continuity of events."[4] He concludes: "We may, therefore, confidently generalize—we may assert without fear of contradiction that all inorganic forms, without exception, have originated by a process of evolution."[5]

Geology and astronomy were still relatively young sciences in the late nineteenth century, with much to learn, but Le Conte felt no trepidation about extrapolating from their findings to bolster his argument for Darwinian evolution. His claim that "all inorganic forms, without exception" evolved gradually would be impossible to justify then or now. Le Conte was in no position to claim that astronomy's infinite universe implied gradualism and evolution, much less that astronomy had it right about the universe being infinite. He was extrapolating into unknown territory, and indeed since Le Conte's day several scientific arguments have arisen against the universe's being infinite. But evolutionists have been fond of this argument, and despite the lack of scientific support, some still make references to the infinite universe.[6]

Le Conte was building a general argument for a materialistic view of natural history. He did so by appealing to the findings of geology and astronomy and comparing them to the doctrine of creation. His discussion followed the pattern of "we once believed . . . , but we now know. . . ." For example:

> There was a time when rocks and soil were supposed to have been always rocks and soils; when soils were regarded as an original clothing made on purpose to hide the rocky nakedness of the new-born earth. God clothed the earth so, and there an end. Now we know that rocks rot down to soils; soils are carried down and deposited as sediments; and sediments reconsolidate as rocks.[7]

It is not clear from which religious tradition he dredged up the notion of soil being regarded as original clothing for the earth, but in any case the argument does very little to support his argument against creation, simply because it is a particular religious view, of which there are many. This is far from the general argument against creation that Le Conte intended.

Having made a general argument for gradualistic evolution in the inorganic world, Le Conte next attempts to make a general argument for why gradualism also prevails in the organic world. He begins with a discussion of the classification problem. There are many different species, both alive today and in the fossil record, and they are difficult to classify neatly into groups. Instead of clustering into distinct groups, the species sometimes run together and fill the spectrum between the groups. Furthermore, different fossil species can be lined up in a sequence suggesting gradual evolution between them.

Small–Scale Change

Le Conte next adds the argument from small-scale change. He cites domestic breeding examples, such as horses, cattle, sheep, dogs, and pigeons, as well as examples from the wild showing that populations can be modified, even to the extent of forming new species. For Le Conte this finding not only suggests gradualism in the organic world but also militates against the doctrine of creation. Le Conte sets up a hypothetical argument with a creationist committed to the idea that small-scale change is impossible. Le Conte places the creationist in an impossible predicament, because small-scale change is observed in domestic breeding experiments and in the wild.

A Naturalistic Bias

The problem, here again, is that Le Conte's argument against a particular religious view does little to support his assertion of gradualism. Certainly his fossil sequence and small-scale change evidences provide possible explanations for evolution. But these must be weighed against all the problems with evolution. Questions such as how life arose in the first place, how complex structures could come about as the result of blind forces, and why we should believe that small-scale change, such as changes in color or size, is evidence for large-scale change such as new organs or added limbs are just a few of the problems that must be considered along with the positive arguments.

Le Conte's evidences are interesting, but they hardly prove evolution or make it compelling. Nonetheless, Le Conte concludes the chapter with the high claim that these evidences make evolution a necessary truth:

> Evolution is certainly a legitimate induction from the facts of biology. But we are prepared to go much further. We are confident that evolution is *absolutely certain*. Not, indeed, evolution as a special theory—Lamarckian, Darwinian, Spencerian—for these are all more or less successful modes

of explaining evolution . . . but evolution as a law of derivation of forms from previous forms; evolution as a law of continuity, as a universal law of becoming. In this sense it is not only certain, it is axiomatic.[8]

Le Conte here makes a leap from a few arguments for the plausibility of evolution to its absolute certainty. It seems that his confidence comes mostly from the arguments against creation, for as he reveals, his confidence stems not from the fossil data or small-scale change but from the assumption of a materialistic view:

> The origins of new phenomena are often obscure, even inexplicable, but we never think to doubt that they have a natural cause; for so to doubt is to doubt the validity of reason, and the rational constitution of Nature. So also, the origins of new organic forms may be obscure or even inexplicable, but we ought not on that account to doubt that they had a natural cause, and came by a natural process; for so to doubt is also to doubt the validity of reason, and the rational constitution of organic Nature. The law of evolution is naught else than the scientific or, indeed, the rational mode of thinking about the origin of things in every department of Nature. . . . The law of evolution is as certain as the law of gravitation. Nay, it is far more certain.[9]

It appears that Le Conte argues for evolution not because it has made successful predictions, as most accepted theories do, but because it is a "necessary truth"[10] according to his presuppositions about the world. The weakness of the positive evidence hardly matters when one already believes that only materialistic explanations are true. And with the acceptance of Darwinism, this materialistic bias became a new truth for evolutionists. It is now a standard justification for evolution, as exemplified in Mark Ridley's textbook:

> Separate creation . . . does not explain adaptation. When the species originated, they must have already been equipped with adaptations for life, because the theory holds that species are fixed in form after their origin. An unabashedly religious version of separate creation would attribute the adaptiveness of living things to the genius of God; but even this does not actually explain the origin of the adaptation, it just pushes the problem back one stage. . . .

> We can accept that an omnipotent, supernatural agent could create well-adapted living things: in that sense the explanation works. However, it has two defects. One is that supernatural explanations for natural phenomena are scientifically useless. The second is that the supernatural Creator is not explanatory. The problem is to explain the existence of adaptation in the world; but the supernatural Creator already possesses this

property. Omnipotent beings are themselves well-designed, adaptively complex, entities. The thing we want to explain has been built into the explanation. Positing a God merely invites the question of how such a highly adaptive and well-designed thing could in its turn have come into existence.[11]

It is little wonder that many people do not believe in evolution. Whether coming from Le Conte in 1888 or Ridley in 1993, these sorts of metaphysical meanderings say more about evolutionists than they do about evolution. Both Le Conte and Ridley argue for the necessity of evolution using unscientific arguments. Le Conte argued that nonnatural explanations were invalid because only natural explanations were rational. Similarly, Ridley finds that nonnatural explanations are "scientifically useless" and "not explanatory" because they don't explain how God came into existence. But Le Conte's and Ridley's premises, that only natural explanations are rational and that God was designed, respectively, are nonscientific. They are statements of personal belief.

Evolutionists have always struggled with the problem of how to promote their theory in the public arena. They believe evolution is a fact, yet according to polls many people are not so taken with the idea. Rarely do evolutionists see that the stumbling block for many is not a lack of understanding but a fundamental disagreement over the presuppositions inherent in the theory. Piles and piles of evidence won't help when it is all based on metaphysical assumptions with which the reader disagrees.

Nonetheless, Le Conte fills his next eight chapters with what he calls "special proofs of evolution." They are needed, according to Le Conte, because although evolution is certain among "scientific men,"[12] it is still not accepted by the popular mind. Le Conte therefore provides the special proofs for people who require such proofs before they can accept evolution as the "only rational mode of thought."[13]

Le Conte's Special Proofs

Unfortunately, what Le Conte calls special proofs are really just his special interpretations. To be swayed by his interpretations one must already believe, as Le Conte does, that evolution is certain. And for those not so swayed, Le Conte consistently reminds the reader that no matter what he or she thinks of the special proofs, evolution is still certain. For example, here is the end of one of his chapters: "In conclusion, let me again impress upon the reader that all the doubt and discussion, above described, as the factors of evolution, is entirely aside from the truth of evolution itself, concerning which there is no difference of opinion among thinkers."[14]

Let's take a brief look at Le Conte's special proofs. One proof involves eight pages of examples of anatomical similarities in different species, with explanations of how some appear to have common origin (homologies) while others have similar function but appear not to have similar origin (analogies). He concludes the proof with the special interpretation that evolutionists commonly make: "Now, one of the strongest proofs of the truth of evolution is taken from the homologies of animal structure. Common origin [i.e., evolution] completely explains homology. Every other explanation is transcendental, and therefore unscientific."[15]

Similarly, Le Conte devotes almost an entire chapter (eighteen pages) to a review of homologies of the vertebrate skeletons before issuing his brief interpretation of such homologies: "The simplest, in fact the only scientific, explanation of the phenomena of vertebrate structure is the idea of a primal vertebrate, modified more and more through successive generations by the necessities of different modes of life."[16] At the end of his chapter on homologies in articulate skeletons, Le Conte pronounces that "the only *natural* explanation, and, therefore the only scientific explanation" for such structures "is that *they were really thus derived* [by evolution]."[17]

The problem with all these "special proofs" is that Le Conte's conclusion does not follow from his premises. He begins by restricting science to naturalistic, nontranscendental explanations; he next argues that evolution is the only naturalistic explanation of homologies. Given these premises, Le Conte correctly concludes that evolution is the only scientific explanation, but he makes the unjustified claim that this is a proof for the *truth* of evolution. Le Conte's unspoken premise is that only naturalistic explanations are true. In assuming this premise Le Conte steps outside the ken of science. His premise is not scientific, for no scientific experiment ever showed this to be true. Ironically, Le Conte is keen to point out the unscientific nature of other explanations but fails to see this in his own explanation.

In addition to his premise that only naturalistic explanations will do, Le Conte often uses the argument that God would never have made the creatures that we find in nature. At the end of his chapter on embryology he discusses the development of teeth in whales and concludes that evolution is the "only conceivable" explanation. Why? Because "if whales were made at once out of hand as we now see them, is it conceivable that these useless teeth would have been given them?"[18] Likewise Le Conte finds that the development patterns in fish reveal "a bungling piece of work"[19] and therefore could not have been created.

Finally, in his chapter on the geographical distribution of species, Le Conte states that if species were created specifically for particular geographical locations, then "each species ought in every case to be perfectly adapted to its own environment, and to no other. But, on the contrary, introduced species often flourish better than in their own country, and better than the natives of their new homes."[20]

Obviously, Le Conte has very specific expectations for his hypothetical Creator. Perfect adaptation, as he defines adaptation, is only one of them. And it is the high specificity of his expectations that make his negative theological arguments problematic. For they are beholden to a specific notion of God, and notions of God, no matter how carefully considered, are outside the realm of science.

H. H. Lane, 1923

H. H. Lane was a professor of zoology at the University of Kansas. He was the son of a devout minister and had served in practically every church office open to a layperson. He had been a member of his church for over thirty years. Lane also had an extensive background in academia. He studied at five universities (DePauw, Indiana, Cornell, Chicago, and Princeton) and had been a professor in four colleges and universities. With this background he felt well qualified to respond to a petition from a group of students asking for a series of lectures on evolution and its implications for Christianity.

Lane's response was complicated but uncompromising. He called for the church to adjust to evolution—a new truth discovered by science. And how could the students be sure of evolution? Lane knew that he had to establish the truth of evolution, and he attempted to do so in a chapter titled "The Fact of Evolution" in his book *Evolution and Christian Faith*, which was based on his lectures.

Species Not Well Adapted

Lane begins with a survey of historical ideas surrounding evolution. He then presents a series of evidences and arguments that he believes firmly establish the fact of evolution. His first argument is that there are many organisms that are poorly adapted. This is, according to Lane, "a hard blow to the advocates of special creation, for it would indicate a lack of skill or foresight not to be thought of in an all-wise and all-powerful Creator."[21]

One problem with this reasoning is that it is difficult to measure adaptation accurately. Even the simplest of organisms have many processes and structures that are still not understood and cannot be monitored continually. Nor is it possible to construct and evaluate alternative versions that might be better adapted. In fact, no one has even come forth with an alternate design for organisms that are thought to be poorly adapted. Today, and even more so in Lane's day, claims of poor adaptation are interesting but remain unjustified.

A more important problem with Lane's reasoning, independent of the problem of measuring adaptation, is that it is really a religious argument in disguise. If we assume for the moment that Lane's assertion of poor adaptation is true, all that he succeeds in proving is that a particular religious view is untenable. He presupposes a certain concept of God—that God must make all species well adapted to their environments—and proceeds to disprove it. What about religious traditions that do not see God's purpose as one of servicing creation? Obviously they would not be swayed by Lane's logic. Job, for example, saw species as created imperfectly by a sovereign God.[22] Nonetheless, evolutionists continue to be fond of using fitness as a universal design criteria that is to be applied to the doctrine of creation just as much as it is applied to evolution. The difference is that if species were created then their fitness should be perfect because the Creator is all-good and all-wise.

Homology

Lane's next argument is from homology—structural patterns that appear across a range of species. Again, his argument is really a religious one, for he argues that "on the basis of special creation [homologies] have no meaning or else seem to limit the exercise of creative power."[23] For Lane, homologies provide strong evidence not for evolution but against creation.

Similarly, Lane moves on to a series of arguments from vestigial organs and embryology. The above discussion on the difficulty in measuring adaptation is relevant here too. In Lane's day many structures were considered to have no function, though similar structures in other species were seen to serve a function. It seemed like clear evidence for evolution. Structures that had become no longer necessary through evolution were not always discarded altogether. In its imperfect process, evolution merely reduced the size of the now useless structure until it no longer interfered with the organism's overall fitness.

Lane's argument from vestigial structures has the same two problems as his argument from imperfect adaptation. First, it is difficult to measure fitness, so it is difficult to say whether a structure really is vestigial or not (see "The Problem of Measuring Fitness" in chapter 2). Indeed, most of the structures Lane would have identified as vestigial are now known to have useful functions. And second, Lane's argument is used primarily as an argument against a particular religious view: "This fact [of vestigial structures] has no meaning on the hypothesis of special creation, while on the hypothesis of descent with modification it finds a *satisfactory explanation*."[24]

Likewise, in his argument from embryology Lane points to the similarities between embryos of different species. In his view God would never have made the species in this way, and he finds this evidence "hard to explain on the basis of special creation."[25]

The Fossil Record

Lane's next category of evidence is the geological record and its fossils. First he finds general evidence in the progression of the fossil record through time. The early record indicates the existence of simple forms, and the successive strata indicate increasing variety and modes of life. This fact speaks conclusively, claims Lane, "against the traditional view of creation" and "clearly in favor of a progressive development."[26]

Certainly this is interesting evidence for evolution, but it hardly rates as conclusive. Lane omits here, for example, the problem that much of the fossil record shows a lack of change rather than "progressive development."

Lane is in no position to make such a strong claim, unless he believes he has come to it by eliminating the alternatives. To be sure, he correctly points out that any and all doctrines of creation that cannot accommodate a progressive fossil record are disproved. The problem is that Lane mistakes this as proof against *all* creation doctrines and therefore an argument for evolution; in fact, he is really only addressing his personal version of "the traditional view of creation."

Lane continues his attack on special creation with a specific example from the fossil record: the horse sequence. In Lane's day the fact that the horse sequence failed to show a smooth, gradual continuum was not as well appreciated as it is today, and so the horse was often touted as obvious and undeniable evidence for evolution. But Lane also saw it as revealing the failure of special creation: "there is not a shadow of evidence anywhere in the whole [horse] series in favor of the hypothesis of special creation."[27]

Lane next appeals to evidence from the geographical distribution of species, which he calls "quite convincing." His discussion consists mainly of an extended quote from Darwin, concluding that the evidences "are intelligible" according to the theory of evolution.[28] It is another example of how often evolution literature relies on mere possibilities.

His next category of evidence is the small-scale change in species achieved by breeders of plants and animals. He concludes the discussion triumphantly: "*This is evolution;* there is involved no hypothesis or theory, in the ordinary acceptation of those terms. It is the *demonstration of a fact* which can no longer be successfully gainsaid."[29]

Finally Lane has arrived at the point where he feels he can speak of evolution as a fact. But what is the fact? A wide variety of small-scale change is possible, but Lane gives no explanation or rationale for why we should believe, contrary to the results of breeders' efforts, that this small-scale change can alter body plans or create new organs.

Ultimately the argument amounts to an equivocation on evolution. On the one hand, evolution is defined as the process that is responsible for all life. Unguided natural forces, acting through the process of evolution, created everything from algae to eagles. But on the other hand, evolution is defined as the mere modification of existing traits—small-scale change.

Early in the twentieth century scientists studied blood immunity and how immune reaction can be used to compare species. Not surprisingly, the blood studies produced results that paralleled more obvious indicators such as body plan. That is, humans are more closely related to apes than to fish or rabbits. Without providing his rationale, Lane claims that this is convincing evidence for evolution.

It is not obvious why the blood immunity data should be chalked up as evidence for evolution. These data can be used to compare and classify species, in the same way visible features had been used in traditional classifications. Comparisons based on different visible features generally agreed with each other, and the blood immunity data showed a similar pattern. The blood data provided a new feature, but conceptually it was not different from the visible features.

Lane's blood immunity argument really amounts to an argument from classification. Darwin had tried to make this argument but ended up making a nonscientific attack on creation that depended on one's view of God.[30] The problem with Lane's assertion is that he fails to give the details of his reasoning. What are the intermediate steps leading to his conclusion? Certainly evolution could explain the blood studies, but, on the other hand, it is not likely that the theory of evolution would have been rejected if the blood studies had turned out differently. Classification work is full of anomalies, and if the immunity data had conflicted with other features, then one of several ad hoc explanations could have been used.

Yes, the species can be classified, and if one presupposes evolution is true, then those classifications can be interpreted as revealing evolutionary relationships. But why should we believe that the classifications serve as noteworthy *evidence for* evolution? Despite these questions, evolutionists have continued to use blood immunity as evidence for evolution. For example, Tim Berra tries to make this argument. He gives a good explanation of the biology involved and how shared similarities and differences are interpreted as revealing evolutionary relationships. But he does not explain how the argument works if evolution is not presupposed.[31] Likewise Edward Dodson and Peter Dodson, without explaining their rationale, argue that only evolution can explain the blood immunity data.[32]

A group of students, curious and thirsty for knowledge, petitioned for a lecture series giving the scientific view of evolution and explaining how they should incorporate it into their worldview. What they got was a series of high claims for evolution that were derived from an unspoken metaphysical position. Lane did give a series of arguments for the plausibility of evolution, but the real strength of his message came from his arguments that God wouldn't have made the world the way we find it.

Arthur W. Lindsey, 1952

Four years after the famous Scopes trial, which found John Scopes guilty of teaching evolution in Tennessee, Arthur W. Lindsey published his *Textbook of Evolution and Genetics*. It includes two chapters on the evidences of evolution. These chapters form the core of the arguments that Lindsey used for evolution in his 1952 book, *Principles of Organic Evolution*. Lindsey was a professor of zoology and biological sciences at Denison University (Granville, Ohio), and he believed that the available evidence made Darwinian evolution "logically conclusive."[33] Nonetheless, Lindsey lamented that outside of science there was still some antagonism toward the idea of evolution, and even where it was accepted it was likely to be regarded as somewhat mysterious.

One problem, according to Lindsey, was that the well-known textbooks in his time dealt mostly with detailed considerations of the factors of evolution and failed to provide a broad survey of evolution. Introductory texts, on the other hand, were too brief to help students develop a comprehensive understanding of evolution. Lindsey intended for his textbook to fill in this gap. It would, he hoped, provide both scientists and nonscientists with the history and supporting evidence of evolution.

Classification and Spontaneous Generation

Lindsey begins his presentation of the evidences of evolution with a discussion of how all life forms are ultimately related to one another. According to Lindsey, relationship itself fails to reveal how the species originated, but he argues that relationship coupled with the failure of spontaneous generation is strong evidence for evolution. The theory of spontaneous generation held that organisms could arise spontaneously. It explained, for example, why rats could always be found in garbage dumps—somehow they spontaneously arose from the garbage. Spontaneous generation was disproved by Louis Pasteur in the nineteenth century, and with its passing biology embraced the law of biogenesis, which stated that all life comes only from preexisting life: *omne vivum ex vivo*—all that is alive came from something living. Since life must come from life, and since all life is related, Lindsey reasons that he has found solid evidence for evolution.

> We have disproved the spontaneous generation of complex organisms so effectively that the origin of all life from preexisting life is regarded as a law of biology, leaving only the initial formation of primordial living substance and possible later syntheses of the same type as links with the inorganic world. The creation of species separately is a philosophic concept without scientific support.[34]

Unfortunately for Lindsey, the argument has less force than he suggests, for the failure of spontaneous generation failed to assist evolution as he had hoped. Yes, life *ordinarily* arises only from life, but it had to begin somewhere. This is the proverbial chicken-or-egg problem, and evolution did not help solve the riddle, for Darwin's process of evolution operates on a preexisting population with variation.

Lindsey's language is imprecise, and his logic has hidden premises. Little would be gained by trying to unpack his argument in order to decipher exactly what he is saying, but it is clear that after all his discussion, Lindsey has fallen back on an argument against divine creation. It is not clear just how Lindsey arrives at his conclusion that creation is to be limited to "a philosophic concept," nor is it clear just what he intends by this terminology. Despite all the confusion, what is clear is that Lindsey had some notion of a doctrine of divine creation and that he has constructed an argument against it which he finds effective in supporting evolution.

Biogeography

Lindsey's next evidence is the geographic distribution of species. He uses an example of moth populations that were distributed from the Mississippi Valley to the Rocky Mountains to explain that a single species can exist in many variants across a geographic range. The different variants shared structure, pattern, color, and habits, and Lindsey wonders why this is so. His answer is that it can be explained only by evolution: "Careful evaluation of these facts leaves only one possible conclusion. They [the different variants] must have been derived from the same ancestry, and as members of the same species this derivation is easily understood."[35]

Lindsey does not explain what he means by "careful evaluation," and the reader is left wondering how evolution is the only "possible conclusion." In fact, evolution fails to explain why small-scale change among moths can be extrapolated to the type of large-scale change that evolution requires. Yes, it is interesting evidence, but without some sort of evidence or explanation of how it can be extrapolated to large-scale change, it remains an unsupported conjecture.

Embryology

Next Lindsey presents evidence from embryology. He uses the circulatory system to explain that the embryos of higher and lower forms, such as mammals and fish, contain similar structures. For Lindsey this is evidence for evolution because the structure has "no known need" in the higher species:

> The circulatory system, especially in the development of the heart and aortic arches, is an equally vivid case. The transition from the two-chambered heart of the fishes to the three chambers of the amphibians and four in reptiles, birds and mammals and the reduction of the aortic arches from the original six pairs to three are significant. . . . In embryonic development . . . the transition is clear. . . . The second pair of aortic arches . . . are supplementary channels which exist for a very short time and *meet no known need. Nature is not in the habit of producing useless structures,* hence we can explain them only as vessels which once were useful and have not been completely eliminated from the developmental sequence.[36]

There are two problems with Lindsey's argument. First, our failure to find a function of a biological system does not imply the absence of any function. Second, even if the absence of function could somehow be proven, Lindsey's conclusion simply doesn't follow from his premise. Clearly the observation that nature is not in the habit of producing useless structures does not permit him to conclude that it *never* does.

Homology

Lindsey's next evidence is from homology. Homologies are similar structures in related species. Lindsey repeats the argument made by Darwin and practically all evolutionists: If God made the species, then there must be no patterns of similar structures—all species must be unique, independent creations that are optimized for their environments:

If special creation were the source of the many kinds of living things, it is reasonable to suppose that each would have the best possible equipment for its mode of life. Instead organisms often have adaptations which definitely resemble those of other species living under very different conditions. One structure may clearly show the essentials of the other, modified to serve a different use. They cannot logically be supposed to have been wholly independent in origin.[37]

Evolutionists make heavy use of this metaphysical constraint, apparently oblivious that it is not scientific and also ineffective for many people. What is "reasonable" for Lindsey to assume about the Creator is unreasonable for many a reader. The problem is that Lindsey is constructing an argument against his particular notion of divine creation and the Creator. Like Darwin's, Lindsey's notion of creation requires that all species be "wholly independent" of each other.

If Lindsey finds evidence for evolution in homologous structures, then what can be said of analogous structures—structures that perform similar function but have very different design? Lindsey considers the eyes of insects, mollusks, and vertebrates, which are all significantly different from each other though they contribute to the same function—vision. Because these species are quite different, evolutionists believe their eyes evolved independently, a process they call convergent evolution. Lindsey concludes that such structures "are excellent examples of analogous organs and convergent evolution"[38] without providing any justification for this remarkable claim. It is true that they are excellent examples *if* evolution is true, but Lindsey fails to explain why they should count as evidence *for* evolution. The title of the chapter is, after all, "Evidences of Evolution."

Lindsey next attempts to generalize the argument from homology. There is, according to him, a great diversity in nature that species do not seem to exploit. From biochemistry to vision, there are many different ways that an organism could perform a given function, yet neighboring species tend to use similar methods, and Lindsey finds this to be strong evidence for evolution:

The *only logical interpretation* available is that the organisms which resemble each other fundamentally have gained their present state as a result of common foundations or, in the terms of evolution, of common ancestral origin. That there are many species which resemble each other in some degree appears to be no more than the result of slightly differing ways of life during the long apprenticeship of their evolution. . . . In short, whether we consider the visible structural relationships or their more elusive chemical foundations, *the organism can build only on the foundations of its heritage,* and only community of heritage can be the source of fundamental relationship in the species that have evolved. . . . If all species were wholly independent in origin, we might expect diversity of structure and function no less than the number of species involved.[39]

Lindsey claims his interpretation is the only "logical" one, yet he relies on the notion that the design of an organism can derive only from "the foundations of its heritage." How can Lindsey justify such a claim? Ultimately it comes down to Lindsey's belief that a Creator would not have created species with commonality—at bottom it is a religious argument.

Adaptation

Lindsey next discusses the ability of organisms to adapt to a wide variety of environments. He uses as an example a remarkable unicellular organism, *Euglena gracilis*, that can perform photosynthesis if sunlight is present but also is mobile and consumes nutrients, such as sugars, if they are available. It could be classified as a plant or an animal. In biology there are many such examples of great variation within the lifetime of an individual. Evolution requires variation, for it is the raw material on which natural selection operates, but natural selection is supposed to operate on variation *between* lifetimes, not *within* a lifetime. Therefore it is not clear how intralifetime variation such as that observed in *Euglena gracilis* relates to evolution. Nonetheless, Lindsey argues that it is evidence for the theory: "The response of *Euglena* to an environment in which sunlight is present and its life in darkness are an excellent example. Just how these varied responses may take part in actual evolution is not yet known, but logically they seem an unavoidable result."[40]

These unicellular organisms are more complicated than a jet airliner, with many processes that we still do not understand. How did evolution supposedly create this? Evolutionists do not know. How could its intralifetime variation participate in evolution? Lindsey does not know this either. But amazingly he felt no trepidation in claiming that it does and that it should be considered as evidence for evolution. Apparently when evolution is assumed to be a fact, just about anything can serve as its evidence.

The Fossil Record

Lindsey's next area of evidence is from the fossil record. The fossil species show long periods of no change, and the appearance of new species is too rapid for the fossils to reveal how they arrived, but the fossil record does document the arrival of new species more advanced than their predecessors. The earlier species are more primitive than those that follow, and for Lindsey this was evidence for evolution:

If a fossil from one epoch is similar to that of the next but more primitive, it is reasonable to conclude that organisms of that kind were undergoing changes at the time, and that the two belong to different levels of development in the same line of descent. For brevity the one may be called ancestral to the other but the approximate relationship is all that can be confidently assumed. . . . Assuming the reality of evolution as a natural process, the most primitive creatures must necessarily have existed in the earliest periods and successive degrees of complexity would have been attained with the passage of time. The general truth of this principle is so evident that the fossil record becomes a further proof of evolution.[41]

Lindsey admits that the actual relationship between species cannot be inferred from the fossil record, but nonetheless he claims it as evidence for evolution. Why? Because it reveals that the more primitive species came first. Ever since Darwin, it has not been unusual for his disciples to think of the evolutionary process as involving increasing complexity, despite the fact that Darwin's theory made no such prediction.

In fact, if we knew of a planet whose life forms never went beyond bacteria, evolution could explain it. Evolution could also explain a hypothetical planet whose life forms progressed only to later regress. On our own planet we have, for example, the trilobites, whose various forms make up a bizarre menagerie, with advanced and unique versions appearing in the early stages. There are also the horselike species, which do not seem to blend into each other but rather remain unchanged until their extinction.

Evolution seems to be able to accommodate all these observations— it is a remarkably flexible theory. But the cost of such flexibility is the loss of confirmation. A theory that boldly predicts a single, distinctive outcome is strongly confirmed when that outcome is experimentally observed. But a theory that accommodates a wide range of outcomes will not be so resoundingly confirmed. Lindsey fails to appreciate this and as a result arrives at an overly optimistic assessment of the fossil evidence. The fossil record may be evidence for evolution, but it is not the strong evidence he would have us believe it to be.

Lindsey set out to make evolution "logically conclusive," but he did nothing of the sort. He laid out a series of arguments that he hoped would convince the reader of evolution, but he ended up either making leaps of logic or sounding more like a theologian than a scientist as he talked about what God does and does not do.

Sir Gavin de Beer, 1964

Sir Gavin de Beer authored several books on evolution, as well as other topics, in his prolific career. He was a scientist, professor, historian, director of the British Natural History Museum, and Darwin scholar. Considering his credentials, it is certainly worth having a look at his arguments for evolution. They appear in his *Atlas of Evolution*, in the chapter "Evolution Is a Fact"—a title that leaves little doubt about de Beer's orthodoxy. Like some of his fellow evolutionists, de Beer not only believed that evolution is a fact but also believed in science as a way to the truth, as he illustrates with this story in the book's preface:

> It is instructive to recall the words of Sir John Pringle in answer to George III when the monarch remonstrated with him at the recommendation made by the Royal Society in favour of a type of lightning conductor devised by Benjamin Franklin, then a mutinous subject in open rebellion against his sovereign. "Sire," said Sir John, "I cannot reverse the laws and operations of nature," thereby establishing the superiority of scientific truth above any other consideration of reason.[42]

De Beer's claim about "the superiority of scientific truth" does not derive from science. Most scientists would say that science must always be tentative, but de Beer speaks of a "scientific truth" and sees it as superior to "any other consideration of reason." And since evolution is supposed to be scientific, it already has the advantage over other explanations—a view that recalls Joseph Le Conte's prescription that only naturalistic explanations can be true. Along with this belief de Beer has a high view of natural laws: "The laws of nature have been found to be of universal application and are held to represent fundamental truths. They are inscrutable and cannot be evaded, suspended, or ignored by a scientist without sacrifice of intellectual integrity."[43]

Homology

Chapters 6 and 7 will say more about how this view of natural law relates to evolution. Suffice it to say for now that de Beer is echoing a pre-Darwinian notion that had an influence on Darwin and the development of his ideas. As for de Beer's evidences for the fact of evolution, he begins with what he considers perhaps the most important evidence: homology. He runs through a number of examples of corresponding structures in different species, such as reptilian and mammalian jaw structures. He argues that the latter evolved from the former through a complex series of changes that included a functional transition where reptilian jawbones ended up contributing to the mammalian hearing system:

> During this functional transition, however, down to the minutest detail, the relations of these little bones to their surrounding structures, nerves,

arteries, and muscles, preserve a plan that is identical with that seen in the reptilian jaw, in spite of the difference in function which the structures perform.[44]

De Beer sees the question of how such intricate changes in these highly complex systems could possibly have come about by unguided genetic variations as no problem. He doesn't even mention it as he concludes that evolution is the only possible explanation: "The facts of comparative anatomy in both vegetable and animal kingdoms are so numerous and cogent that even if there were no other approaches to the problem, they would suffice to prove that only evolution can account for them."[45]

For de Beer, homologies are not just evidence for evolution; they are *proof of* evolution. Similarly, Darwin had stated that if no other evidence were available, homologies alone would be sufficient to convince him that evolution is true. Such confidence ignores the unresolved issues that come along with the homology argument, such as the fact that there is actually no objective, unambiguous method for identifying which similarities are homologies and which are analogies. But de Beer makes it clear with a Darwin quote that his confidence comes from the metaphysical interpretation of homologies that is so popular with evolutionists: "Nothing can be more hopeless than to attempt to explain this similarity of pattern in members of the same class, by utility or by the doctrine of final causes."[46]

According to de Beer, evolution can account for homologies, but competing explanations cannot; therefore *only* evolution can account for them. The proof comes by the process of elimination, and a metaphysical idea was implicit: the doctrine of final causes cannot account for homologies.

Embryology

De Beer's next category of evidence is embryology. Animals pass through various embryonic stages as they develop from a single cell—the zygote—to the adult form. Because all animals start life as a zygote, they all resemble each other at that stage, no matter how different they may be in the adult stage. A fish and a bird begin life in a similar way. As the zygote multiplies and the young animal passes through the embryonic stages, it gradually becomes more distinctive, taking on characteristics particular to its species. But during this process there are stages where the embryos of very different species—a fish and a bird, for example—may yet have much in common.

Ever since Darwin, evolutionists have taken these similarities to be evidence for evolution. They have especially looked for similarities between species that are supposed to be evolutionarily related, for if such a relationship exists then one might expect certain similarities to persist in their developmental pattern. Evolution modifies the organism with an eye toward reproduction, but it doesn't necessarily modify the whole organism. And those early stages of life—the embryonic stages—seem to reveal the most fundamental or basic nature of the organism. Since different species are similar in these stages, evolutionists reason that they are related.

But this is as far as the argument goes. It may appeal to the evolutionist's intuition, but it lacks any details about precisely why the evidence should be counted in favor of evolution. De Beer simply states that this is the case:

> Nobody would mistake adult lizards for birds or mammals, and if their embryos are so very similar there must be a reason for it. The reason, which Darwin was the first to point out, is that all animals whose embryonic stages are similar are related and descended from a common ancestor, from which they have inherited the form of the embryonic stage which they repeat in their own development. . . . The similarity between embryonic stages is evidence that evolution has occurred.[47]

We can agree with de Beer that there must be a reason the embryos of lizards and birds are similar, but from there de Beer gives no rationale for his conclusion that these similarities are "evidence that evolution has occurred." What we need is an explanation of why these observations point to evolution. The evidence would be more convincing if it could be shown that with evolution only these types of embryos would have come about. But evolution makes no such bold prediction; rather, it accommodates a wide range of possibilities.

Vestigial Organs

If de Beer is impressed with the evidence from embryology, he is even more impressed with the evidence from vestigial organs—those structures that, evolutionists say, have lost their usefulness and are declining in size and importance. Like most evolutionists, de Beer mistakenly assumes that such structures have been proven to have no function. He is confident in his assessment of function, or lack thereof, and for him this is "confirmation of evolution."[48] Why? Because the presence of such useless structures "is inexplicable"[49] except on the view of evolution. Divine creation *cannot* explain them.

Such a view of course depends on one's concept of creation. De Beer makes his view clear in his use of a quote from Darwin on flightless birds:

"There is no greater anomaly of nature than a bird that cannot fly."[50] But why was this considered such an anomaly? Nature has always been known to be full of variety.

Nowadays, with evolution firmly implanted, it may be difficult to see just how nonscientific Darwin's pithy remark is, but it tells much of how the modern age molded the nineteenth and twentieth centuries' view of creation. Nature and its species were expected to be efficient and optimal, according to our understanding of these virtues. Birds were meant to fly, and there was something wrong with those that couldn't. Though Darwin and his peers did not understand nature's inner workings, they were bold in their pronouncements about what virtues nature should and should not exhibit. And nature's failure to fulfill our ideals and expectations was considered clear proof of evolution. All birds should fly, but since some don't, there must be a crude law of nature rather than a Creator behind such incompetence. All structures should have an obvious function; otherwise they must be "vestigial," the result of a random unconscious process.

Some structures that are assumed to be vestigial according to evolution have a different function from their purported ancestral structures. For example, de Beer cites insect wings that must have evolved into gyroscopic navigational devices and muscles that must have evolved into electric organs for signaling or attacking. This calls for a certain amount of serendipity: the blind, unguided process of natural selection is supposed to remake an outmoded structure into something highly complex and quite useful. The upshot is that evolution accommodates all possibilities. Whether a function is discovered or not, the structure can still be safely labeled as vestigial, and as de Beer concludes, even the useful ones "are evidence for evolution."[51]

Biochemistry

De Beer next finds evidence for evolution in the biochemistry of organisms. Using several examples, he makes the point that, in addition to the traditional methods of classification, biochemistry can be used to classify species. The biochemical similarities and differences in everything from cell wall composition to blood proteins "are not scattered at random" but can be classified into groups. The resulting groups often agree with traditional classification, but there are some interesting differences. For de Beer this is yet another impressive proof of evolution. Regarding the cellular components of different species de Beer writes: "When a number of organisms resemble each other and differ from others in the chemical substances which their protoplasm contains, the reason is that they have inherited the ability to form such substances from a common ancestor."[52]

Similarly, de Beer discusses coloration in butterfly wings. In a certain group of butterflies the color is caused by a chemical peculiar to that group. These butterflies must therefore, de Beer concludes without explaining his reasoning, "have inherited this chemical trait from a common ancestor."[53]

De Beer also finds blood protein comparisons between species to supply convincing evidence. These data confirmed the distinction that had been suggested between rabbits and rats, and the data indicated no close affinity between frogs and newts. Furthermore, whales were seen to be closer to even-toed ungulates (ox, sheep, giraffe, camel, pig) than to other orders of mammals; seals were seen to be related to dogs; the musk ox was more closely related to sheep and goats than to cattle; and the panda was closer to bears than to raccoons. These results generally agree with those of traditional comparative anatomy, and for de Beer such a correspondence is "proof of evolution."[54]

Like H. H. Lane, de Beer fails to provide the reasoning behind his conclusions. What are the intermediate steps and premises in these proofs? If the panda is anatomically closer to the bear than to the raccoon, then shouldn't we expect its biochemistry *also* to be closer to the bear's? Nonetheless, de Beer is quite impressed with these evidences. His lack of explanation is exceeded only by the confidence he has in his conclusions.

Parasitism

De Beer's next area of evidence is parasitism. He gives several examples illustrating the various modes of sustenance and propagation in parasites. These can be quite complicated, such as in the case of the liver fluke parasite which de Beer describes. Like the brainworm parasite discussed in chapter 4, the liver-fluke undergoes a remarkably intricate life cycle.

One might think that evolutionists would be at a loss to explain how such complex life histories could be created by the blind forces of evolution. And indeed there are no detailed explanations of how this could happen, but de Beer nonetheless believes that here he has found evidence for evolution. His reasoning is that parasites must have evolved because the host evolved:

> As is the case with any organisms showing marked adaptation, the parasite mode of life cannot have been the original condition of the parasites, which must be descended from organisms that led a normal free-living existence. The structural features of parasites must therefore have been produced during descent with modification from normal ancestors. The

very existence of parasites is evidence that they have become what they are from a previous condition in which they were free-living, and parasites are therefore living proofs of their own evolution.[55]

His logic fails to support his conclusion. The problem lies in his premise that parasites must have evolved from free-living organisms because the parasitic mode of life could not have worked at first since there were no hosts. This premise is certainly reasonable if evolution is true, but de Beer begs the question when he uses it as evidence *for* evolution. If anything, de Beer's argument points out the weakness of evolution in failing to explain how such complex cycles could come about by its unguided process of change.

De Beer also finds evidence for evolution in the host specificity of certain parasites. While some parasites are widely tolerant of conditions and host types, others can survive only in a limited range of conditions and so are restricted to one type of host. Furthermore, related species of parasites are often found to be restricted to related species of hosts. This finding is not surprising and does not need evolution to explain it, but again de Beer sees only an evolutionary interpretation: "These facts mean that related hosts harbouring related parasites are descended by evolution from a common ancestor, which harboured the common ancestor of the parasites."[56]

It seems de Beer is so taken with evolution that he is unable to find anything but supporting evidence and proofs, no matter what observations are at hand.

Classification

De Beer's next evidence for evolution is from the area of classification. This area of biology, as much as any other, has historically been rife with metaphysical interpretation. And it continues to be so with evolution. De Beer gives a lengthy classification example, but his main point is a metaphysical one, and he makes it with a Darwin quote:

> The several subordinate groups in any class cannot be ranked in a single file, but seem clustered round points, and these round other points, and so on in almost endless cycles. If species had been independently created, no explanation would have been possible of this kind of classification.[57]

It has been known since Aristotle that species tend to cluster in a hierarchical pattern, and in the eighteenth century Linnaeus saw it as a reflection of the Creator's divine plan. Obviously this pattern does not force one to embrace evolution. Also, Darwin's law of natural selection does not predict this pattern. Darwin had to devise a special explanation—his *principle of divergence*—to fit this striking pattern into his overall theory.[58] To be sure, evolution can accommodate this hierarchical pattern, but the pattern is not necessarily implied by evolution. From Darwin to de Beer, the real strength of the argument for evolutionists, over against Linnaeus, has been in its use against divine creation. "If species are separately created," wrote evolutionist George Carter, "there is no reason why they should be created in large groups of fundamentally similar structure."[59]

More recently, paleontologist Niles Eldredge has elevated this argument, making the remarkable claim that the hierarchical pattern found in the classification of species really is a necessary consequence of evolution.[60] Eldredge uses an analogy of monks in the Middle Ages copying manuscripts to support his bold assertion and concludes that biology's hierarchical pattern, far from being fitted into the theory of evolution, instead now should be viewed as a test that evolution has successfully passed.[61] Eldredge has high confidence in the classification data—more, it seems, than is warranted by his analogy with medieval monks. It could be that his confidence is bolstered by evolution's metaphysical interpretation of the evidence. Like Darwin, de Beer, and others, Eldredge repeats the refrain that these hierarchies would never have been designed by an intelligent Creator: "Could the single artisan, who has no one but himself from whom to steal designs, possibly be the explanation for why the Creator fashioned life in a hierarchical fashion—why, for example, reptiles, amphibians, mammals, and birds all share the same limb structure?"[62]

Eldredge has tried to elevate the classification data to the status of a successful prediction, but like the other evolutionists he also uses the data against divine creation. The evidence disallows divine creation—at least the kind of creation that evolutionists have in view. Darwin, de Beer, and the others could not conceive of a God who would have his species so grouped, so evolution must be true.

Biogeography

De Beer's next category of evidence is biogeography, or the geographical distribution of species around the globe—a fascinating but, as it turns out, complex and subtle area of research. Darwin enlisted biogeography to support evolution, as did his contemporary Alfred Wallace, who studied the area in depth for many years. The Darwin-Wallace paradigm dominated research over the next century, although many have argued this was due more to the acceptance of evolution than to the success of the paradigm. In fact, the Darwin-Wallace model has been criticized for having led to more anomalies than firm conclusions and to more imaginative scenarios than testable hypotheses.[63]

De Beer was writing just before the Darwin-Wallace model was questioned, so it is not surprising that he cites biogeography as evidence for evolution. After his time it is less common to find biogeography cited so unequivocally as evidence for evolution.[64] Although biogeographical patterns clearly support the idea that species undergo small-scale changes, they no longer serve as obvious evidence for the large-scale change that evolution requires. With the acceptance of plate tectonic theory, new theories of biogeography emerged that were not necessarily committed to evolution.[65]

What does remain true is that attempts to marshal biogeographical evidence for evolution, from Darwin to today, typically rely on arguments against the doctrine of creation. Alec Panchen, for example, agrees that building an argument for evolution from biogeography is not easy, but he believes it is possible. He constructs an argument that includes, as its second proposition, the opinion that "it is improbable that the distribution of organisms can be explained by the separate creation of species [because] ecological adaptation in any environment is demonstrably imperfect."[66] Panchen's unspoken assumption here is that under divine creation all species must be perfectly adapted—a nonscientific claim.

Though biogeography no longer provides the clear support for Darwin's evolution as was once thought, the built-in metaphysical premises remain interesting for our story. Consider the Darwin quote that de Beer selects:

> Shall we then allow that the distinct species of rhinoceros which separately inhabit Java and Sumatra and the neighbouring mainland of Malacca were created, male and female, out of the inorganic materials of these countries? Without any adequate cause, as far as our reason serves, shall we say that they were merely, from living near each other, created very like each other . . . ? Shall we say that without any apparent cause they were created on the same generic type with the ancient wooly rhinoceros of Siberia and of the other species which formerly inhabited the same main division of the world; that they were created less and less-closely related, but still with interbranching affinities, with all the other living and extinct Mammalia; that without any apparent adequate cause their short necks should contain the same number of vertebrae with the giraffe; that their thick legs should be built on the same plan with those of the antelope, of the mouse, of the monkey, of the wing of the bat, and of the fin of the porpoise; . . . that in the jaws of each when dissected young there should exist small teeth which never come to the surface? That in possessing these useless abortive teeth, and in other characters, these three rhinoceroses in their embryonic state should much more closely resemble other mammalia than they do when mature. And lastly, that in a still earlier period of life, their arteries should run and branch as in a fish, to carry blood to gills, which do not exist. . . . I repeat, shall we then say that a pair, or a gravid female, of each of these three species of rhinoceros, were separately created with deceptive appearances of true relationship,

with the stamp of inutility of some parts, and of conversion in other parts, out of the inorganic elements of Java, Sumatra and Malacca? Or have they descended, like our domestic races, from the same parent stock? For my own part I could no more admit the former proposition than I could admit that the planets move in their courses, and that a stone falls to the ground, not through the intervention on the secondary and appointed law of gravity, but from the direct volition of the Creator.[67]

Darwin packs more metaphysical thought into this short passage than can be found in most church sermons. He uses his metaphysics to find the doctrine of divine creation faulty on several counts. Are we really to believe, he asks, that the rhinoceros was created from inorganic materials? And can we believe that, "without any adequate cause," rhinoceroses would be created with the "stamp of inutility" revealed by so many homologies?

This remarkable quote could be the subject of an entire chapter, but what is perhaps most interesting for our purposes is Darwin's concluding thought. After tallying up all the problems with his straw-man version of creation, Darwin appeals to the high view of natural law that, as we shall see, was quite prevalent in his day. It was safe for Darwin to use gravity as an analogy, because everyone believed that there was no direct act of providence in gravity. Likewise, Darwin argues, we should not believe that there is a Creator directly involved in the biological world. Of course no one can say whether or not a divine hand is behind every action of gravity, and Darwin's presumption that there is none says much about the metaphysics behind evolution.

The Fossil Record

De Beer's final area of evidence is paleontology—the fossil remains of plants and animals. This is a lengthy section filled with many photographs and illustrations, and the majority of the discussion deals with the supposed process of evolution rather than the evidence for evolution. Regarding the actual evidence, de Beer makes one argument, which he splits into three parts. The argument is that species in the fossil record can be arranged in a sequence—what de Beer calls a succession of forms—where certain anatomical features reveal a trend.

A prominent example is the horse sequence, where the species became larger and the number of toes decreased over time. De Beer also cites species that can be placed between different classes, the so-called transitional species, and he cites the refinement of certain sequences as the other parts of the argument. De Beer's point is that although the fossil record does not provide a complete spectrum of species within a sequence, it does provide occasional discrete snapshots of the process as it went along.

Unfortunately this argument was, and remains today, speculative. Did these different horse species really evolve one into the other? As we saw in chapter 4, instead of one species gradually blending into the next improved version, the fossils reveal a spectrum of discrete varieties. New species coexist with older ones, and species seem to emerge rapidly.

If there are evolutionary relationships between the different specimens, we cannot safely infer them from the fossil record. It would be just as reasonable to model the horse sequence simply as different species. De Beer, however, props up his argument for evolution with a rather unabashed swipe at divine creation. He states that the successive forms in the fossil record are evidence of change, and unless "one is prepared to believe in successive acts of creation and successive catastrophes resulting in their obliteration, there is already a strong presumptive indication that evolution has occurred."[68] In other words, de Beer cannot believe that God would have done it that way, so evolution is already the obvious choice.

Verne Grant, 1991

Verne Grant is professor emeritus of botany at the University of Texas. He has contributed to the field of evolution for forty years; in his ninth book, *The Evolutionary Process*, he presents a brief three-page summary of the evidence for evolution. Grant wants to keep his book readable and concise, and he succeeds in this section. He enumerates eight areas of evidence, each with a paragraph or so of explanation, which he believes together form a powerful argument for the truth of evolution:

1. direct observation
2. extrapolation to larger groups
3. fossil record
4. taxonomic pattern of relationships among living species
5. geographical distribution
6. homology
7. vestigial organs
8. biochemical similarities

In this survey we have already seen nonscientific arguments about why these evidences are supposed to compel us to believe in evolution. Grant does not use reasoning significantly different from that of the other authors. But this itself is worth noting: in Grant the reasoning that evolution is true because the doctrine of divine creation is false now seems to be a tradition. He needs only to conclude each paragraph with a terse phrase or sentence, or maybe a quote from Darwin, to remind the reader of what is already common knowledge. For example, in category 4 (taxonomic pattern of relationships among living species), Grant concludes that "living species would not be expected to cluster in groups within groups if they were products of separate acts of creation."[69] In this case Grant undermines his own conclusion by acknowledging that this evidence was known and accepted by preevolutionary taxonomists who incorporated it into their creationist views.

In category 5 Grant concludes that "the doctrine of creation provides no explanation for the observed patterns of geographical distribution of supraspecific groups."[70] For category 6 he uses a famous Darwin quote about nothing being "more hopeless than to attempt to explain" homologies by creationist schemes. In category 7 Grant concludes that "there is no good explanation for the existence of useless rudimentary organs in the doctrine of creationism."[71] He concludes discussion of category 8 by noting that biochemical similarities provide the same evidence as category 4.

Categories 1 and 2 deal with small-scale change. Category 1 is small-scale change itself, and category 2 is the justification for why small-scale change must be extrapolated to the large-scale change that evolution requires. Grant believes small-scale change explains the large-scale change because there is a fundamental unity of nature that we must not violate by positing two different mechanisms: "It is inconceivable that organic diversity should be produced by [small-scale] evolution at the lower levels but by some other means at supraspecific levels. Separate hypotheses to account for the origin of low-level and high-level organic groups are unwarranted."[72]

Grant does not just argue that the small-scale change could possibly be extrapolated to large-scale change; he argues that such extrapolation is necessary. But he relies on a metaphysical premise to make his argument work. Though nature is full of surprises and often uses unique and distinctive mechanisms, Grant requires that nature use the same mechanism for small- and large-scale change, though he has no scientific justification for such a claim. Finally, in category 3 he makes the argument, similar to de Beer, that the succession of forms in the fossil record is evidence for evolution.

A Pattern of Claims

This survey shows that evolutionists who have attempted to rigorously prove their theory have routinely resorted to nonscientific claims.

There are plenty more examples of this sort of thinking beyond the five works examined here. Paul Moody argues that if scientists have not discovered a useful function for an organ such as the appendix, we should assume there is no function. He then argues that only a Creator who lacked skill in planning or construction would create such functionless organs.[73] And for Michael Ruse, God cannot be reconciled with the facts of biogeography, so we must turn to evolution. He argues, "Given an all-wise God, just why is it that different forms appear in similar climates, whereas the same forms appear in different climates? It is all pointless without evolution."[74] According to Edward Dodson and Peter Dodson, if God created the species, then they should be distributed uniformly about the globe. They write, "Had all species been created in the places where they now exist, then amphibian and terrestrial mammals should be as frequent on oceanic islands as on comparable continental areas. Certainly, terrestrial mammals should have been created on these islands as frequently as were bats."[75] It is remarkable how often evolutionists feel free to dictate what God should and shouldn't do.

In chapters 2 through 4 and now in chapter 5, we have seen a long series of remarkable arguments for evolution. Evolution, like gravity, is supposed to be a fact, yet the arguments put forth in support of it are either arguments for the mere plausibility of evolution or arguments against the doctrine of divine creation. Over and over we find arguments about why God wouldn't have done it that way, which work only with a certain concept of God.

But these arguments have held sway now for almost a century and a half. Why? The answer to this question lies in the popular understanding of God. For though evolutionists make extensive use of a certain concept of God, they did not contrive that concept for their own convenience. It was, rather, formed over many centuries, long before Darwin was ever born.

6

Modernism before Darwin

I n theology, the doctrine of God is the basic doctrine on which the other doctrines are built. Likewise, secular thought is often guided by an underlying view of God. Charles Darwin often made reference to the Creator, as did his early followers and today's evolutionists. Their references to God are not simply for the purpose of historical context or cultural perspective but are part of the basic reasoning. Evolution needs these metaphysical interpretations of scientific observations. Nonetheless, evolutionists don't seem to feel the need to justify their religious assumptions.

How did the evolutionist's notion of God become so popular that it needed no justification? The answer lies in the history of religious thought. The details of this story are beyond the scope of this work, but there was a general trend, especially in the modern age, toward a comprehensible deity. God's acts of creation and providence were increasingly being approached as things that could be understood, reasoned about, and even scientifically modeled. In the extreme, God was becoming more a theory than a person.

These sorts of ideas can be found in a variety of movements in the centuries leading up to Darwin's time. Rationalism, the Enlightenment, deism, the nonconformists, and Unitarianism were all different movements contributing to this trend. For some thinkers, doctrines such as the fall and the Trinity were mysterious and unnecessary. More important for our story, the idea that God would use direct intervention or miracles was increasingly questioned in favor of the idea that God acts exclusively via natural laws. After all, modern science had found that the motions of everything from planets to apples are governed by the same simple laws. Perhaps all phenomena—even such things as the flood and God's moral laws—could be explained by natural laws.

But while the findings of science were influential, science itself was not the motivation for these movements. Most of these ideas were fueled by speculation rather than by empirical observation. As nineteenth-century historian W. E. H. Lecky wrote: "Few of the grounds upon which the more serious skepticism of the nineteenth century is based then existed. One of the most remarkable differences between eighteenth century Deism and modern freethinking is the almost entire absence in the former of arguments derived from the discoveries of physical science."[1] Indeed science was, if anything, often seen as a ready defense *against* skepticism.[2]

Two recurring ideas have fueled the notion that God acts according to natural laws: *divine sanction* and *intellectual necessity*. In the former, God is seen as being all the greater for designing a world that works on its own rather than requiring his divine intervention. In the latter, the restricting of God to natural laws is urged because only this ensures that meaningful scientific inquiry is possible. If natural laws are liable to violation, then we cannot discern the law from the exception.

In the centuries preceding Darwin, philosophers and scientists increasingly scrutinized God according to the rules of reason and sought to describe creation as operating according to natural laws. Among other things, they were trying to explain creation and its evils. What sort of natural schemes did God use to bring about the world, and how can evil be explained? For if God is powerful and loving—omnipotent and omnibenevolent—then where and how did evil arise? Evil is divided into two broad categories: moral and natural. Moral evil is perpetrated by human beings and is usually understood as the negative side effect of the autonomy that God wanted his creatures to have. But natural evil has no such easy explanation—death by lightning rather than a sword cannot be ascribed to a sinful heart. Natural evil can include everything from hurricanes to snakebites, and it raises questions about the very nature of creation. Various thinkers attempted to explain how God made the world and why natural evil is part of it, but none of the solutions were very satisfactory.

Darwin too was working out a solution to the problem of natural evil. His effort to explain the existence of evil (this time in the form of carnage and inefficiency) is apparent in his notebooks and his book on evolution. And his solution to the problem of natural evil fell right alongside those of previous thinkers—not because he was necessarily directly influenced by them but because they all were working from a similar notion of the Creator and the creation. Darwin's great contribution to this tradition was the strong scientific flavor he gave to the solution, to the point that most readers lost sight of the embedded metaphysical presuppositions. Whereas the earlier solutions lacked detailed explanations, Darwin provided scientific laws and biological details.

But Darwin's general approach, to distance God from creation by interposing natural laws, followed the earlier attempts. Looking back now, with almost two centuries of hindsight, we see that Darwin's use of natural selection operating in an unguided fashion on natural biological diversity was a unique solution. But his overall approach, to distance God from evil, was predictable.

Creation

Since the Middle Ages, nature has often been viewed as having a certain amount of autonomy. God may have arranged things in the beginning, but since then nature has, for the most part, operated according to uniform, eternal natural laws—a process unto itself. For some this added to God's glory, for the Creator must be all the more wise to construct a universe that operates on its own, with no need of divine adjustment.

Creation Explained Scientifically

An example of how the high view of natural laws influenced early scientists is the cosmogonies of the seventeenth century and later that tried to explain how the world came about. They were sometimes constructed so as to explain the biblical account, including the fall and the flood. But the biblical events were not viewed as the result of a direct divine cause. Rather, they were described in terms of the actions of natural laws. God was a passive Creator, as the world would be explained as the result of natural laws. It didn't seem to matter that there was precious little scientific observation from which to work. These thinkers felt they could reason their way to the truth, because creation was believed to operate via eternally fixed natural laws.

René Descartes (1596–1650) was the first of the great rational philosophers and for many the first modern philosopher. Descartes was interested in nature and creation. He searched for processes, based on hypothesized laws and interaction between matter, that would create our complex world from a simpler, more homogenous state. He presented a mechanistic cosmogony using vortices to create planets and stars. Descartes's particular solution would not survive the eighteenth century's applications of Newton's laws, but his general approach of searching for mechanistic explanations would. The early Darwinist Thomas H. Huxley would later acknowledge evolution's debt to the Cartesian worldview.[3]

Thomas Burnet, Edmund Halley, and William Whiston all followed Descartes with mechanistic cosmogonies. Burnet (1635–1715) authored the most popular geologic work of the seventeenth century, *Telluris Theoria Sacra (The Sacred Theory of the Earth)*, in 1681. Here he tried to provide a geological rationale for all biblical events. He saw the earth as evolving from a chaotic primeval void to a perfectly ordered sphere. Included in this evolution was the unfortunate but required destruction of its own paradise. But as natural laws and moral laws are interconnected, it was simultaneous with the fall. Thus Burnet provided naturalistic explanations for biblical events. God could cause a flood, but he would do so by secondary means. Burnet described why this view of the Creator was in keeping with his faith:

> We think him a better Artist that makes a Clock that strikes regularly at every hour from the Springs and Wheels which he puts in the work, than he that hath so made his Clock that he must put his finger to it every hour to make it strike; And if one should contrive a piece of clock-work so that it should beat all the hours and make all its motions regularly for such a time, and that time being come, upon a signal given, or a Spring toucht, it should of its own accord fall all to pieces; would not this be look'd upon as a piece of greater Art, than if the Workman came to that time prefixt, and with a great hammer beat it into pieces?[4]

In other words, we should expect God to use secondary means because this requires all the more wisdom and foresight. A God who simply exercises direct control over his creation is less impressive. Here Burnet gives us an early example of the divine sanction for a less than sovereign God who is distanced from his creation. God may be almighty, but he is all the more impressive because he does not exercise his might. Instead of exercising brute force, he governs exclusively through secondary means.

The great astronomer Edmund Halley (1656–1742) proposed that a comet, rather than God's hand, could have caused the biblical flood.[5]

William Whiston (1667–1752), Isaac Newton's successor at Cambridge, elaborated on Halley's idea and proposed that the earth could have been created by a comet and the flood caused by yet another comet. There were others who put forth cosmogonies in addition to Descartes, Burnet, Whiston, and Halley, but all of them reflected and promoted the high view of nature's laws and uniformity.

The world, it seemed, was an ongoing, self-sufficient creation machine. Everything was assumed to be driven by God's natural laws. The German philosopher Gottfried Leibniz (1646–1716) extended this notion even to God's disciplinary actions. Leibniz envisioned a system in which natural law literally leads to moral law. He believed the universe was made up of an infinity of monads—simple substances—assembled in a hierarchy from the smallest particles up to God. The universe was a huge machine where punishment for sins comes from the very operation of nature. Morality was built into the laws of creation. The psalmist says that God renders to each according to his or her work.[6] For Leibniz this was the result of nature's mechanism: "sins carry their punishment with them by the order of nature and by virtue of the mechanical structure of things itself; and . . . in the same way noble actions will attract their rewards by ways which are mechanical as far as bodies are concerned, although this cannot and should not always happen immediately."[7] It was modernism's own version of "you will reap what you sow,"[8] achieved by autopilot.

Hume's Rejection of Miracles and Uniformitarianism

This view of a Creator who does not dabble in his creation was also evident in David Hume's attack on miracles. Hume (1711–1776) made a persuasive argument against the very possibility of miracles and reversed them from being an apologetic to a liability for believers. He began with the observation that a miracle is something that violates natural law. Otherwise it is not a miracle. Therefore when one is presented with a miracle claim, it becomes a matter of "proof against proof." One must weigh the evidence for the natural law versus that of its breaking. There are the many years of observations that support the law versus the single instance of its breaking. Obviously, the proof of the single instance would need to be highly compelling to overcome the many observations of the law in action. According to Hume, the appropriate criterion to apply to a report of a miracle is that it ought to be considered true only if its falsehood must be more miraculous than the reported miracle itself. Hume was as much historian as philosopher, and in looking over the various miracle claims of history, he found none that provided the necessary proof.

Later Hume's argument was found to be problematic. C. S. Lewis, for example, claimed Hume argued in a circle, for he must have presupposed nature to be absolutely uniform.[9] Hume's proof has problems, but what is important for our purpose is that the flaw was not perceived in Hume's day, so strong was the tug of science and its high view of natural laws. "The more we know of the fixed laws of nature," Darwin would later write, "the more incredible do miracles become."[10]

Consider the case of Johannes Kepler's formulating his mathematical relationships of planetary motion given Tycho Brahe's series of observations. By careful study of many observations Kepler found that planets travel in elliptical paths. Were Kepler's results eternal and unchanging laws or merely nominal laws, liable to not yet observed suspension? With a high view of nature's laws and uniformity, modern thinkers found Kepler's laws to be eternal rather than nominal; to admit exceptions would be to introduce gratuitous contrivances.

But in carefully avoiding gratuitous contrivances, science goes too far. It not only grants no contrivances within the observed data; it also grants none outside the time-and-space window of observation. In fact, Brahe's observations do not allow us to declare Kepler's laws eternal and unchanging, nor do the observations allow us to presume that the laws can be occasionally suspended. Both explanations are equally wedded to their respective metaphysics.

In most cases, science's presuppositions of simplicity and uniformity are of little consequence and are in fact transparent to the practicing scientist. Hume's proof against miracles is important because it highlights these unspoken assumptions. The proof fails because of these assumptions, but its acceptance reveals how strongly these assumptions had been internalized.

Some have seen Hume's argument against miracles as a precursor to critical history and the problem of evaluating the claims of history.[11] The idea is that, after Hume, historians increasingly rejected the supernatural and placed it outside of history. For example, the eminent twentieth-century theologian Rudolf Bultmann argued:

> The historical method includes the presupposition that history is a unity in the sense of a closed continuum of effects in which individual events are connected by the succession of cause and effect. . . . This closedness means that the continuum of historical happenings cannot be rent by the interference of supernatural transcendent powers and that therefore there is no "miracle" in this sense of the word. Such a miracle would be an event whose cause did not lie within history.[12]

In other words, while we may wish to see God's hand in a historical event, the historian must not view the event as having been caused by

God's hand in any direct sense. In the next chapter we shall see a parallel movement in the uniformitarian urgings of Hutton and Lyell, which called for the Earth's history to constitute a long sequence of uninterrupted natural events. In many ways uniformitarianism assisted Darwin. "Everything in nature," Darwin would confidently state, "is the result of natural law."[13]

The Problem of Evil

As with creation, one's view of evil is profoundly influenced by one's view of God. The problem of evil has been with us since antiquity. It asks how an omnipotent and omnibenevolent God could allow evil to exist in the world. Apparently God is either unable to control the evil, and is therefore not all-powerful, or else he can control the evil but chooses not to, in which case he is not all-good.

Many solutions, or theodicies, for the problem of evil hold that evil arose from the autonomy that God granted to his creation. God installed natural laws between himself and creation, and evil was somehow an unfortunate byproduct of the workings of those natural laws. God is distanced from creation and is therefore absolved of its evil. The problem of evil could be solved with careful explanations of how God relinquished control of creation.

A Theodicy for Moral Evil

This sort of theodicy has often been applied to the problem of moral evil. In this case, moral evil arises from the fallen state of the human subjects. It is not that God cannot control evil but that he places a high premium on the autonomy of the creature, so God allows for the existence of evil.

This solution reached a literary pinnacle in one of Charles Darwin's favorite works, John Milton's *Paradise Lost* (1667). *Paradise Lost* is considered one of the great works of literature. It was immensely popular in Darwin's day, and Milton remains a hero of the faith for many today. Milton's stated objective was to "justify the ways of God to man." His ambitious tale traces the cosmic plot of Satan's disobedience, the fall of humanity, and God's offer of salvation.

In Milton's view God created the world and everything in it, so a link between Milton and Darwin may seem unlikely. But Darwin's evolution says less about the creation act than it does about God's providence after the initial creation. Some see a foreshadowing of Darwin's evolution in Milton's verse; for example, Milton describes creatures not as fixed and unalterable but rather as having potential for growth.[14] But there is a more fundamental link with evolution. In Milton's view of things, God never caused evil to occur. When God made matter it was perfect, and it became imperfect after God loosened his hold on it. Here Milton deviated from the Scriptures he knew so well. "I bring prosperity," the Lord declared through the ancient prophet, "and create disaster."[15] But for Milton, God should be distanced from the world's evil ways. It was not that God is surprised by historical events, but he is, in a sense, disconnected from them. God knows the future before it happens, but this is because he is omniscient, not because he influences the future. This pious view of God protects him from evil, but it also makes the world somewhat independent of the Creator.[16]

For Milton, creation was originally perfect yet mutable, susceptible to evil after God relinquished control. Likewise for Darwin, the world began as "laws impressed on matter by the Creator," with subsequent "production and extinction . . . due to secondary causes"[17] outside of God's control. The point here is not that Darwin was directly influenced by Milton, although he may have been, but rather that there is a parallel between Milton's and Darwin's views of God and evil. Both sought a rational concept of God that they could incorporate into their work; for example, both Milton and Darwin doubted the mysterious doctrine of the Trinity. And both men distanced their God from evil. Milton, and many thinkers after him, believed God's creative acts must only have produced perfection, as we define perfection. The evil we find in the world must somehow have arisen on its own. Likewise, Darwin relied on the view that God can only create harmony and efficiency. Biology's quandaries must have arisen by natural law.

Theodicies for Natural Evil

Milton's type of theodicy has been and continues to be popular, but he applied it to moral evil. What about natural evils such as disease, pestilence, and earthquakes? As modern science began to unlock the secrets of the natural world, there came increased curiosity about its failures and why God allowed them. Milton's theodicy used the strategy of distancing God from moral evil, and several attempts were made to distance God from natural evil as well. Modern science's general approach was to characterize the natural world in terms of natural laws, so an obvious strategy, and ultimately the one used by Darwin, was to use the natural laws themselves as the source of natural evil.

About the same time that Milton wrote *Paradise Lost*, a small group of Anglican divines known as the Cambridge Platonists[18] began to explore the problem of natural evil. When Jesus healed a blind man, the disciples wondered whether the cause of the blindness was the man's sin or his parents' sin. Jesus explained that neither was the cause—the man was blind so that the works of God could be revealed in him. In other words, this instance of natural evil could not be explained by rational analysis. But the Cambridge Platonists did seek rational explanations. They spoke of the "errors" and "bungles" in nature and argued that the universe developed gradually in the absence of divine guidance. They distanced God from creation with the notion of a "plastic nature," where a spiritual deputy of God directs the natural processes.[19] As in Milton's solution to moral evil, the cause of natural evil was not the Creator; instead, evil came about as an unfortunate byproduct of a mechanism external to God.

Milton and the Cambridge Platonists were not intending to offer alternatives to the Scriptures; rather, they were trying to elucidate the Bible's creation story. The Cambridge Platonists were not the first to attempt this sort of theodicy. The ancient Gnostics also tried to explain how a good God could have created such a messy creation. Their idea of semidivine rulers is similar to the Cambridge Platonists' spiritual deputy:

> Most Christian gnostics made some attempt to avoid an overt dualism, realizing that a complete disconnection between God and world could not be accepted by the Church. Their task, therefore, became that of making an indirect connection which would nevertheless exonerate the good God of all guilt in regard to this world. Often the connection was made through a series of archons (semidivine rulers) who reigned over the lower spheres in the absence of God. The more archons and spheres existed between the high heaven and the cosmos, the greater the distance between God and creation, the less God's guilt for having allowed this calamity to happen.[20]

Just as the ancient Gnostics had found that matter is evil, spirit is good, and the two must be separated, so too the Cambridge Platonists found that nature contains evils from which God must be distanced. But the idea of a plastic nature was rather vague, and thinkers would continue to deliberate on the problem of natural evil. The Cambridge Platonists are important for our story because within the modern age they are an early example of the Creator's being distanced from creation to explain natural evil.

Leibniz's Theodicy

A more sophisticated response to the problem of natural evil came from Gottfried Leibniz. Isaac Newton's laws of motion explained how celestial bodies move about their paths, but for Newton they also pointed out how necessary the Creator is, both for the initial arrangement and for subsequent maintenance and adjustment of the system.[21] For Leibniz this cosmic tinkering posed a dilemma, for if God intervenes against his own laws, he contradicts his own creation. At the very least, this would be the sign of an imperfect craftsman. A better explanation for cosmic anomalies is that the universe logically had to be imperfect, otherwise it would not be distinct from God. That is, the creation could not, by definition, possess the absolute perfection of the Creator and not be a part of the Creator. If God was to create anything at all, it would have to be imperfect. Imperfection—or evil—is simply a natural part of creation. Thus Leibniz rescued God from Newton's clumsy providence and simultaneously produced a theodicy that could explain natural as well as moral evil.

It may be true that creation must be less than perfect, but how much less? It seems that our world is terribly evil—much more so than is required by Leibniz's logic. Surely God could have created a world with fewer earthquakes, fires, and floods. For this objection, Leibniz pictured the world as a complex machine. Yes, a world with less evil was certainly possible, but there would be much less good as well. The objective was not to minimize evil but to create a world with much more good than evil. It was a design tradeoff, and God used the best design possible. We might say God maximized the good-to-evil ratio.[22]

Implicit in Leibniz's theodicy is that the criterion of goodness resides outside of God. When God made the world, it was very good not because God made it but because God was a good engineer. God did not determine virtue; rather, God worked according to virtue.

Like Milton, Leibniz justified God by distancing him from evil, even at the cost of a less than almighty God. The Creator was now less personal and more of a theory. Leibniz did not believe that there was a divine hand involved in the course of history, for this would mean that God tinkers with his creation. The providence of God was a result of God's wisdom and foresight rather than his active role in history.

Such views were largely shared by English physician and scientist Nehemiah Grew (1641–1712). Grew believed, for example, that the miracle stories were real but not directly supernatural. Yes, the walls of Jericho fell, but they did so as a result of an earthquake. There was nothing supernatural, for it was a result of natural law in action. The miracle is in the precise timing, which only God could have arranged by building it into the laws of nature from the beginning. Nature itself, with all of its laws, is one continuous miracle. It follows that Grew did not believe that God was active in the creation. Grew reasoned, with Leibniz, Burnet, and others, that such active providence would reflect a rather amateurish Creator and that we should not expect God to be the immediate cause of anything in nature. God was constrained to act through his natural laws.

A century and a half later Darwin would continue the tradition. For example, Darwin lamented that providentialists do not appreciate how God's magnificent laws are capable of producing "every effect of every kind which surrounds us."[23]

As for natural evil, Grew blended Leibniz's theodicy with a healthy dose of utilitarianism, anticipating later naturalists such as William Paley. Grew wrote:

> The water flows, the wind blows, the rain falls, the sun shines, heaven and earth act and move, and all plants live and grow for the use and benefit of sensible creatures, and all inferior creatures for the service of those above them. Nor is there any one of so many parts which compose every creature but what is either necessary for its being or convenient for its better being. As it has nothing hurtful or redundant, so no agreeable part is wanting to it.[24]

Grew had a rather idealistic view of the natural world. He argued that apparent defects serve a greater good. Birth defects, for example, help us to appreciate healthy children. But Grew's was a rather glib treatment of evil. Surely there were plenty of evils in nature that would not yield to such facile explanations. Leibniz too did not seem to appreciate the extent of evil in the world when he argued that the world's evil is overshadowed by all the good. Furthermore, he provided no rigorous or detailed explanation of why we should believe that this particular world we live in has been optimized to maximize the ratio of good to evil. Leibniz had to agree that one could certainly imagine worlds with much less evil, but his argument that they must contain much less good was based more on faith than anything else.

In the end this theodicy is not very satisfying, especially for those who are sensitive to evil. Leibniz may have justified evil in theory, but what about when it really happens? Can devastating earthquakes, such as the one that shook Lisbon in 1755, really be contributing to the overall good? In 1759 the great French author Voltaire criticized this apparently credulous optimism in his philosophical tale *Candide*. Voltaire poked fun at Leibniz with his character Dr. Pangloss, who ludicrously saw everything working out for the best in the world.

Hume's Argument against Natural Theology

The problem of evil was also important in the work of Hume. The skeptical philosopher showed how the existence of evil undercuts natural theology—the use of nature to supply proofs of God. Natural theology had been popular for centuries, and for many it had become a justification for faith, sufficient to prove the existence of God. But Hume argued that nature's secrets do not always fit so easily into natural theology's mold. Nature is sometimes more confusing than comforting, and if it can be used to prove God, then it can also be used to disprove God—or at least push him ever further into the background. The reversal of natural theology is fair game, for the original premise was that God's very existence is open to mortal scrutiny. If we can confirm the divine, then we can deny the divine as well.

Hume argued that natural theology was living well beyond its means. Evidence for design can be used only to infer some sort of process sufficient to produce what is observed, and no more. Hume pointed out that when natural theologians argue for the God of the Bible, they are inferring more from their data than they were entitled. Indeed, Hume's character Philo turned the argument around by pointing to the evil in the world. How is that to be accounted for? Hume asked. If God is omnipotent and omnibenevolent, then there should be no evil.[25]

Darwin was also concerned with the problem of natural evil as he developed the theory of evolution. There is an obvious parallel between Hume and Darwin. Both overturned long-standing traditions (natural theology and divine creation, respectively) that explained the world as a result of God's creative activities. And both Hume and Darwin based their arguments on the existence of natural evil.

From Milton to Leibniz, Hume, and others, modern intellectuals were rapidly moving away from the view that God creates and controls the world and toward the view that God must be separated from evil. In the nineteenth century these views would play an important role in Darwin's development of evolution. The common denominator between Darwin's evolution and the earlier theodicies is that God governs via secondary causes—his fixed natural laws—and that God is justified to humankind when we view natural evil as a result of some sort of cosmic constraint outside of God. Darwin worked within this tradition, and it is no surprise that he arrived at his theory of evolution, which claims that nature's imperfections and evils arose from natural forces rather than a divine hand.

Darwin followed this tradition but filled in the details. Where Leibniz could not provide the fine points of his system, Darwin provided overwhelming detail. The theodicy of separation would finally be complete—but not without consequence. God was finally rationalized, but it seemed also that he was lost altogether. For many this conclusion had the sanction of modern science, but underlying the technical jargon was a God who had to be distanced from the apparent failings of his creation.

7

The Victorians

Often our fundamental assumptions about reality are transparent to us. They are like metaphysical spectacles which influence how we see the world, though we have forgotten we are wearing them. This is the case for the theory of evolution. Its metaphysical spectacles, shaped centuries earlier, had become so comfortable that by the nineteenth century few thinkers were conscious of them. Yet evolution claims to be objective, free of any nonscientific premises. There is a disparity between evolution's claims of objectivity and its use of metaphysics.

In the nineteenth century, the opinion among intellectuals that God was superfluous in philosophy and science grew from a minority position to the consensus. One might think it was a time of remarkable change, but there was a silent thread of constancy that ran throughout this great transition. Though God became unnecessary, the popular concept of God remained basically intact. Darwinism led to a God who was not necessary, and at the end of the century Friedrich Nietzsche declared that God was dead. For Nietzsche humanity was finally free of God, but what was less obvious was that humanity was not free of the religious premises on which the movement was built. One cannot disprove God without first assuming something about God. Humans may have been free of God, but they were not free of their presuppositions about God.

Presuppositions about the nature of God were as strong as ever in the nineteenth century, but the dramatic removal of God from the scene led many to conclude falsely that metaphysics had been dropped altogether. Many assumed the metaphysical spectacles had finally come off, when actually they were becoming all the more comfortable—the metaphysical spectacles had become metaphysical contact lenses. With God removed from the picture, the myth arose that metaphysics had been completely routed. Biology seemed no longer beholden to any religious presuppositions. The arguments against divine creation that led to evolution in the first place were now mistakenly viewed as merely side arguments—polemics against a now discredited religious belief. Evolutionists claimed to be free of any metaphysical ties, and they would never look back.

How was it that the removal of God could be construed as the removal of metaphysics? The key is that the modern doctrine of God had become strongly internalized. No longer was it a *particular* doctrine of God; rather, it had become *the* doctrine of God. Therefore the removal of modernism's God was seen not as the removal of a particular type of theology but rather as the removal of God altogether. This chapter examines how the modern doctrine of God influenced early nineteenth-century thought and Darwin's formulation of evolution.

Rational Theism

There was a great diversity of religious belief in Darwin's time, ranging from romanticism and existentialism to evangelicalism and revivalism. One common thread running through most of this diversity was a decidedly human-centered outlook. Romanticism emphasized personal experience over against God's revealed truth. Evangelicalism also focused on personal experience, this time in pursuit of moral improvement. Even the Oxford Movement in the Church of England, reacting against the creeping liberalism and rationalism accommodated by the church, was a return not so much to a sovereign God as to ecclesiastical authority, ritual, and liturgical formality. People were religious, but they tended to focus on themselves more than on God.[1] It was the first time in Europe, A. N. Wilson concludes, that "a generation was coming to birth who had no God, or no God of any substance."[2]

Amidst this milieu of religious thought, two important themes are discernible in the writings of Darwin and his fellow naturalists: Gnosticism and natural theology. Gnosticism is an ancient belief system that draws a strong distinction between spirit and matter. Spirit is good and matter is evil. Whereas the Bible says that God made the world, Gnosticism holds that God is separate from the world; thus Gnosticism is a theodicy. Yes, there is evil, but it is far from God. God is separate and distinct from the world and not responsible for its evils. In Darwin's time the world was increasingly seen as controlled by natural laws. God may have instituted these laws in the beginning, but he had not since interfered; the laws were now his secondary causes. As in Gnosticism, God was seen as separate from the world.

Since God was separate from the world, natural phenomena were not interpreted as results of divine providence. This view seemed to have a divine sanction; after all, to control the world exclusively through natural laws—God's secondary causes—required an even greater God. In other words, a clean separation of God and creation made for an even purer God, just as the Gnostics had found that spirit could be good when it was opposed to matter. In 1794 Darwin's grandfather Erasmus Darwin wrote this Gnostic-sounding statement of how natural history should be viewed:

> The world itself might have been generated, rather than created; that is, it might have been gradually produced from very small beginnings, increasing by the activity of its inherent principles, rather than by a sudden evolution by the whole by the Almighty fiat. What a magnificent idea of the infinite power of the great architect! The Cause of Causes! Parent of Parents! Ens Entium! For if we may compare infinities, it would seem to require a greater infinity of power to cause the causes of effects, than to cause the effects themselves.[3]

The ancient Gnostics were also antihistorical. Whereas the Bible presents a history of God's activity in the world, including dates and historical figures, the Gnostics believed that God's revelation was not open but secret—revealed from within rather than in public documents such as Scripture. Furthermore, whereas the Bible says that the heavens declare the glory of God, the Gnostics believed that one should not look for signs of God in nature.[4] In Darwin's day, a parallel view developed that urged the separation of religion and science; this view remains strong today (see chapter 8 for further discussion).

A striking example of the Gnostic tendencies of Darwin's time arose when John Millais's painting *Christ in the House of His Parents* was first exhibited at the Royal Academy in 1850. In the painting, the boy Jesus has injured his hand in his father's carpentry shop. Mother Mary attends to the boy while Joseph continues with his work; outside the door sheep patiently await their future Savior. The scene is both symbolic and realistic, with wood scraps scattered on the ground and workers going about their duties. But the scene was altogether too realistic for a generation whose God had become abstract and spiritualized. The Scriptures say that God became flesh and lived among us.[5] He knew sorrow, pain, temptation, and joy. But this view of God was lost on the Victorians; they emphasized God's wisdom, power, and transcendence. Could God really have bruised his hand in a messy carpenter's shop? *The Times* complained that the painting was revolting: its "attempt to associate the holy family with the meanest details of a carpenter's shop, with no conceivable omission of misery, of dirt, even of disease, all finished with the same loathsome meticulousness, is disgusting." *Blackwood's Magazine* said, "We can hardly imagine anything more ugly, graceless and unpleasant," and Charles Dickens called the painting "mean, odious, revolting and repulsive."[6]

The Victorians could not believe that Christ the Savior could become involved with creation any more than the Gnostics could. As one historian of Gnosticism put it, "If Christ is to be taken seriously as the Savior how can he actually be part and parcel of this material cosmos?"[7] The Gnostics could not believe God became a man for the same reasons they could not believe God directly created the world—they could not envision God involved in a world so fraught with misery. Similarly, just as the Victorians were troubled by Millais's depiction of the human side of Jesus, they would have trouble with the idea that God created the biological world, apparently so full of inefficiencies, anomalies, and useless bloodshed.

Natural theology was another common theme in the work of naturalists of Darwin's time. In contrast to the Gnostic theme, natural theology does look for signs of the Creator in his creation. Nature displays complexities and intricacies that surely must be manifestations of an intelligent designer. But while Victorian natural theology ascribed nature's wonders to the Creator, it tended to avoid nature's quandaries. The evil side of nature was either ignored or redefined as something positive.

In Darwin's day natural theology was most often associated with the work of William Paley (1743–1805). From 1790 to 1802 Paley produced three works that used classic natural theological arguments, albeit updated to reflect the knowledge of modern science. If the design and crafting of a modern instrument such as a watch requires our careful, determined labor, how much more does the human body show us how necessary a Creator is? This is Paley's famous "watchmaker" argument. The strength of his exposition should not be underestimated, but he presented an overly optimistic view of the world, for he failed to account for its evil. Nature's complexity implies an intelligent designer, but what about pain and suffering?

Controversy later arose over Paley's reliance on utilitarianism—the notion that happiness should be maximized—but his general approach to natural theology remained popular with later naturalists such as Adam Sedgwick. At the foundation of Paley's system was his belief "that God wills and wishes the happiness of his creatures."[8] The Scriptures speak of creation's groaning, but Paley's reasoning would lead one to expect harmony as a necessary outcome in God's world. What happens when it eludes us?

The early Darwin had read and was impressed with Paley's arguments,[9] but Darwin would later see the evil in nature that had bothered Hume. Like Hume, Darwin attempted to account for natural evil. If God would have harmony, then chaos and destruction imply no God. In this way Paley inadvertently helped set up Darwin for his theory of evolution.

There were many others who, like Paley, urged a happy view of nature. For eighteenth-century naturalist Griffith Hughes, creatures he observed were "without defect, without superfluity, exactly fitted and enabled to answer the various purposes of their Creator, to minister to the delight and service of man, and to contribute to the beauty and harmony of the universal system."[10] In the nineteenth century Darwin's friend Charles Lyell also saw in nature a wise Creator at work. "Whatever direction we pursue our researches," wrote Lyell in 1830, "whether in time or space, we discover everywhere the clear proofs of a Creative Intelligence, and of His foresight, wisdom, and power."[11] Job's foolish ostrich[12] would have no place in these idyllic versions of nature.

Gnosticism and natural theology were common themes in Darwin's day, and they were built on a rather pleasant concept of God. This was the sort of God that Darwin had in view as he developed his theory of evolution. Philosopher Michael Ruse has observed that

> Darwin was obviously no traditional Christian, believing in an immanent God who intervenes constantly in His creation. Most accurately, perhaps, Darwin is characterized as one who held to some kind of "deistic" belief in a God who works at a distance through unbroken law: having set the world in motion, God now sits back and does nothing.[13]

But Darwin was not alone here; few if any of his fellow naturalists believed that God constantly intervenes in creation. Nor did Darwin and his peers believe that the world should ever offend our sensibilities. Those naturalists ranged from orthodox to liberal in their religious beliefs, but *rational theism* was a common thread between them. Creation's uniformity, harmony, and symmetry were emphasized over its particulars and anomalies, and God's role was more to service creation than to execute his sovereign plan.

Miracles Viewed As Clumsy

Rational theism also meant for many that God would not interfere in his creation with miracles. No less than the lord chancellor of England, the outspoken and prodigious Henry Peter Brougham (1778–1868), commented on the subject. Brougham entered high school at the age of seven, the University of Edinburgh at the age of fourteen, and found practically no subject beyond his grasp. He was simultaneously an abolitionist, advocate for the people, and defender of the throne. His successful defense of Queen Caroline in 1820 against an attack from Parliament brought Brougham to the pinnacle of fame. For him the physical sciences were a hobby, and he issued his own edition of Paley's *Natural Theology* in 1835. For Lord Brougham, miracles proved nothing but the exercise of power, and they left the Creator's trustworthiness in question. Jesus entreated Philip to believe at least for the sake of the miracles,[14] but for Brougham harmony rather than the supernatural argued for God.[15] Scottish social reformer George Combe (1788–1858) also suspected that the picture of a continually interfering God was flawed. He argued that God governs through unchangeable laws and not supernatural interventions.[16]

Brougham's and Combe's views are samples of the early nineteenth-century view against miracles, which recalls Burnet more than Hume. Whereas Hume had calculated a logical proof for the impossibility of miracles, they now were becoming, perhaps more important, distasteful. Feeding the multitudes, healing the sick, and resurrecting the dead were once obvious signs of God's grace, but they now were increasingly viewed as awkward anomalies unbefitting a truly wise and powerful God.

The Works of a Pleasant God Revealed

The *Bridgewater Treatises* provide yet more examples of this view of a benevolent, wise, and aloof Creator. The *Treatises* assembled eight eminent scientists, by to the will of the Earl of Bridgewater who died in 1829, for the purpose of demonstrating "the Power, Wisdom, and Goodness of God, as manifested in the Creation."[17] This purpose statement alone is an insight into the era. The focus was on how the pleasant aspects of God could be seen in creation. It is not that the gruesome aspects of nature were completely ignored, but when they were considered, they were force-fitted into the happier view of God.

One author, for example, translated nature's death and bloodshed, which after all is relatively swift and painless, into a divine "dispensation of benevolence."[18] He argued that modern science reveals the "infinite wisdom and power and goodness of the Creator."[19] Another author was able to find signs of a happy God in even the most extreme examples of parasitic behavior.[20] Where the apostle Paul might have seen creation groaning,[21] the Victorians could find signs of a pleasant Creator. Rarely was nature used to prove God's wrath or justice.

In 1826 Lord Brougham argued that science affords us "an understanding of the infinite wisdom and goodness which the Creator has displayed in all his works."[22] The influential American naturalist James Dwight Dana wrote in 1856 that God's successive creations of species are revealed in the fossil record "in their full perfection."[23] This was a consistent theme for many scientists in the early and mid-nineteenth century. There was much talk of the Creator and his creation, but the link between them was restricted. God's goodness and wisdom were thought to be manifest in creation, but not his providence, judgment, or use of evil.

Uniformitarianism

We saw in chapter 6 that David Hume's argument against miracles helped promote the view of human history as a continuum of natural events rather than being susceptible to one-time supernatural events. In natural history this notion can be seen in uniformitarianism, belief in a stable creation with fixed, predictable laws. Its best-known exponent was Charles Lyell (1797–1875), a lawyer and scientist who advanced its cause in the 1830s. Lyell's forerunner was James Hutton (1726–1797), who believed earth's internal heat energy was the source of geological change. Hutton saw change as gradual and uniform. His premise was the inviolable regularity of natural laws,[24] and he described the earth as a self-renewing "beautiful machine." He concluded his work with the memorable and metaphysically laden phrase "We find no vestige of a beginning—no prospect of an end."[25] This was a rallying cry for uniformitarianism if there ever was one, for how else could science describe the past except by eternal laws?

Lyell postulated that no geological forces have ever acted besides those now acting, in their present manner and degree.[26] "Let us suppose," he submitted, "that the laws which regulate the subterranean forces are constant and uniform, (which we are entitled to assume, until some convincing proofs can be adduced to the contrary)."[27] Lyell could not *prove* that natural laws were the sole agents of change, but he nonetheless claimed that uniformitarianism should be the default paradigm—assumed to be true unless and until proven false. The burden of proof was shifted to the skeptic.

Lyell's arguments derived not so much from any new discoveries as from a thorough review of available information, cast into compelling arguments for uniformitarianism. Its antithesis, catastrophism, had persisted for many years and not without reason. Catastrophism had its supporting evidence,[28] so one of Lyell's key lines of argument was the imperfection of geologic data. The strata provide snapshots, not a continuum of earth's natural history. They were, in Lyell's view, isolated data points that required extrapolation to form the big picture, and the notion of a gradually evolving earth—uniformitarianism—would supply the missing pieces.

According to Lyell, if the geological data are ever ambiguous or confusing to us (such as gaps in the geological strata), it is likely that we are simply ignorant of nature's ongoing phenomena. It is not likely that different phenomena operated in the past. In other words, uniformitarianism will provide us with the correct interpretations, not catastrophism.

Lyell may or may not be correct in this assertion, but we may rightfully ask what scientific experiment led to this conclusion. Lyell had none, but this seemed not to matter, for until uniformitarianism is embraced "the principles of science must always remain unsettled."[29] It was not that uniformitarianism had to be true; rather, uniformitarianism was required for science to properly advance. For Lyell, science made uniformitarianism intellectually necessary. Lyell's uniformitarianism really amounts to faith in a particular worldview, and he states as much: "The philosopher at last becomes convinced of the undeviating uniformity of secondary causes, and *guided by his faith in this principle*, he determines the probability of accounts transmitted to him of former occurrences."[30]

But along with his presuppositions, Lyell advanced thorough and compelling arguments for his thesis. His work was highly influential, and he is considered by most as the father of modern geology. Lyell was a significant influence on Darwin, although this is ironic since Lyell denied transmutation—the origin of species by natural means. Instead, he supplied Darwin with several foundational, if less tangible, building blocks for his framework. Most obviously, Lyell's precedent allowed Darwin to deweight the fossil evidence. If Lyell could argue for uniformitarianism despite the bumpiness of the geological record, then Darwin could argue for evolution despite the abruptness of the fossil record. Also, Lyell's gradualism was amenable to the sorts of mechanisms that Darwin appealed to for organic evolution. Catastrophism would make such evolution less plausible, and, perhaps more important, it suggested that God, rather than natural law, guided the creation process. Finally, Lyell's forms of argumentation proved useful in evolution. As we shall see in chapter 8, Darwin and his followers have appealed to intellectual necessity to reinforce evolution just as Lyell used it to help justify uniformitarianism. Lyell's technique of placing the burden of proof on the opponent also proved useful for evolutionists. As we saw in chapter 4, this was Darwin's approach to the problem of complexity.

Earlier Theories of Evolution

Lyell was not the only thinker who depended on natural law to describe his findings. A plethora of such accounts emerged from various fields. Auguste Comte, founder of positivism, developed sociology primarily as a tool for investigating the laws of social evolution. He found that civilization progressed according to his famous law of three successive stages: the theological, the metaphysical, and the positive.[31] Humankind was now entering the final stage, where superstitions would be laid aside in favor of the truth of reality.

George Combe saw laws in both the natural and the social spheres. The physical sciences had showed, he said, that God governs through eternal laws and not supernatural interventions. There are equally general moral laws governing the political, social, and economic worlds.[32] Likewise, Herbert Spencer posited laws governing human interaction that would ultimately guide us to our complete happiness. Humankind has undergone, and continues to undergo, modifications resulting "from a law underlying the whole organic creation."[33]

Laws were also proposed to explain the origin and diversification of life. In the eighteenth century Comte de Buffon led a tradition of evolutionary speculation based on the Cartesian mechanistic approach applied to the origin of species. Buffon asked the leading question: "Are all the species of animals the same now that they were originally? . . . Have not the feeble species been destroyed by the stronger . . . ? Does not a race, like a mixed species, proceed from an anomalous individual which forms the original stock?"[34]

Buffon influenced such men as Étienne Geoffroy St.-Hilaire and Jean Lamarck, who proposed evolution by the law of inheritance of acquired characteristics. In England Darwin's own grandfather, Erasmus Darwin, in his book *Zoonomia* (1794), argued for the mutability of species. In the nineteenth century Auguste Comte impressed the young Darwin. For Comte, it was the task of social physics to uncover how humanity had progressed from the primitive to the civilized stage. Although Comte's focus was on social, not biological, problems, he found in biology a useful analogy:

> If one concedes that all possible organisms were successively placed during a suitable time in all imaginable environments, most of these organisms would necessarily end by disappearing, thus leaving alive only those that could satisfy the general laws of that fundamental equilibrium: it is probably through a succession of analogous eliminations that the biological harmony was established little by little on our planet, where indeed we still see it modifying itself unceasingly in a similar manner.[35]

Another pre-Darwinian entry in the field of evolutionary thought was *Vestiges of the Natural History of Creation,* written by Robert Chambers but published anonymously in 1844. Chambers saw uniform laws acting in morality and social behavior as well as the physical sciences. The inorganic world "has one final comprehensive law, Gravitation. The organic . . . rests in like manner on one law, and that is,—Development."[36] Thus Chambers argued for evolution by natural means. The work was not scientific and was full of unfounded speculation, but nonetheless was widely read and aroused much controversy.

Naturalistic ideas for the origin of the species were in the air, but no one had succeeded in setting forth a specific and scientifically acceptable hypothesis. What was needed was a natural law, or an aggregate of laws—a process. Astronomer and philosopher John Herschel outlined the problem at hand in 1836: "The origination of fresh species, could it ever come under our cognizant, would be found to be a natural, in contradistinction to miraculous, process—although we perceive no indications of any process actually in progress which is likely to issue in such a result."[37]

Darwin was inspired by Herschel's statement. True, Herschel was pessimistic about science's chances of finding a solution to the speciation problem, but Darwin would handle that part of the problem. What was important for Darwin was that the subject was being raised and the prospect of a scientific solution was being considered.

William Whewell, in his *History and Philosophy of the Inductive Sciences* (1837), raised the possibility of a law-based scientific theory of creation. Like Herschel, he was none too confident of success: "Nothing has been pointed out in the existing order of things which has any analogy or resemblance, of any valid kind, to that creative energy which must be exerted in the production of a new species."

Whewell did not believe that science would be successful. The process of speciation was clearly not currently acting, and thus science would be required to peer into the past and accurately detect such a process. For this reason, science could not impinge on Scripture regarding the beginning of things. Nonetheless, Whewell, like many nineteenth-century thinkers, had a healthy respect for scientific claims. If a theory of transmutation could be proved, then Scripture must be reinterpreted:

> When a scientific theory, irreconcilable with its ancient interpretation [Scripture], is clearly proved, we must give up the interpretation, and seek some new mode of understanding the passage in question, by means of which it may be consistent with what we know; for if it be not, our conception of the things so described is no longer consistent with itself.[38]

In other words, while science has not yet solved the problem of development, it must be heeded if it someday succeeds with a solution. This essentially amounted to a dare. If science can come up with a successful theory, then so be it—our interpretation of Scripture must be revised.

The Problem of Evil

Something of a foundation had been laid for Darwin. There was Hutton's and Lyell's uniformitarianism that justified interpolation through the fossil-record gaps and Lyell's argument from intellectual necessity for naturalistic explanations in science. Then there were various evolutionary theories, which, if nothing else, inspired pundits to ponder the possibility of such a theory.

Evil versus Morality

None of the previous theories succeeded in establishing a law or process that scientifically described the origin of species. Darwin developed such a law, but in the years before he published *Origin*, not all thinkers believed a law of speciation was even possible. For while it was permissible to use natural laws for just about every phenomenon under study, the one exception for some thinkers was the origin of species. This was a dividing line that some would not cross. There was no apparent dilemma if God uses secondary means to create and guide the inorganic world around us, but what if life itself is the result of mechanical action?

Although Kant had destroyed the theistic proofs of natural theology, he also concluded that the moral law within is both necessary and sufficient for the proof of God. It would make no sense to have the law of morality in conflict with the law of speciation. If the law of morality made God necessary, then the law of speciation should not make him unnecesary. In what we might call the *problem of morality*, how can God be so aloof from his creation, allowing even his creatures to be the result of blind mechanical forces, yet simultaneously be the source of our moral law and the ultimate judge of our actions? In the former he has become nonexistent, or at least irrelevant; in the latter he is vital. For some, a law of speciation would endanger the moral law.

So the battle lines were already drawn before Darwin ever published his theory. Thinkers such as Herbert Spencer, Robert Chambers, and Baden Powell had embraced the notion of evolution by law. But others such as Charles Lyell, Adam Sedgwick, and Louis Agassiz had already rejected the possibility. Lyell, staunch supporter of natural law and mentor to Darwin, would take almost a decade to accept evolution after the publication of *Origin*.

Modern Examples of Natural Evil

But where the problem of morality mandated a divine presence in the world, the problem of evil needed a divine absence. Thinkers have for thousands of years recognized this problem. Aristotle observed that evil is more plentiful than good and that what is hateful is more plentiful than what is fair. A disenchantment with nature contributed to Epicurean materialism, and Pliny declared, "God, if God there be, is outside the world and could not be expected to care for it." Closer to the nineteenth century, Hume had used nature's hostility and destructiveness to argue against natural theology.[39]

The Victorians were, however, unearthing new levels of detail on the quandaries of nature. To begin with, as new species from around the world continued to be discovered, naturalists tried to classify them according to God's hierarchy. This problem, inherited from the eighteenth century, was not becoming any easier. As different species were compared to determine how they should be arranged, a dizzying network of relationships was found. For example, George Shaw (1751–1813) published in 1799 his finding in southeast Australia and Tasmania of the remarkable duck-billed platypus.[40] The platypus has enough characteristics to classify it as a mammal, yet, unlike other mammals, it is oviparous (it produces eggs that hatch outside the maternal body). So how should the platypus be classified? The growing number of known species and their intertwined relationships called for a Creator who must have committed an inordinate amount of effort to the mundane details of nature.

Another problem with nature came from Thomas Malthus and his 1798 essay on population dynamics, which seemingly contradicted rational theism's wise and benevolent deity. Malthus used the authority of mathematical formulation to explore the problem of exponentially growing populations that outpace available resources. The seemingly inevitable result: a downturn caused by starvation or disease. Had God made more mouths than he could feed, and was nature now not trustworthy if natural populations underwent wild swings? The Bridgewater authors put forth feeble attempts, based on utilitarianism, to allay such fears. They consistently interpreted Malthus's population swings as an example of God's wise providence, but such explanations were strained.

Natural Evil in Verse

The nineteenth-century poet Alfred, Lord Tennyson was also concerned with biological evils. Mourning the death of a friend in his famous *In Memoriam*, he found nature "red in tooth and claw." God and nature seemed to be at odds, for nature was indifferent to the loss of life and even whole species.

> The wish, that of the living whole
> No life may fail beyond the grave,
> Derives it not from what we have
> The likest God within the soul?
>
> *Are God and Nature then at strife,*
> That Nature lends such evil dreams?
> So careful of the type she seems,
> So careless of the single life;
>
> That I, considering everywhere
> Her secret meaning in her deeds,
> *And finding that of fifty seeds*
> *She often brings but one to bear . . .*
>
> "So careful of the type?" but no.
> From scarpèd cliff and quarried stone
> *She cries, "A thousand types are gone:*
> *I care for nothing, all shall go.*
>
> "Thou makest thine appeal to me.
> I bring to life, I bring to death:
> The spirit does but mean the breath.
> I know no more." And he, shall he,
>
> Man, her last work, who seemed so fair,
> Such splendid purpose in his eyes,
> Who rolled the psalm to wintry skies,
> Who built him fanes of fruitless prayer,

> Who trusted God was love indeed
> And love Creation's final law—
> *Tho' Nature, red in tooth and claw*
> *With ravine, shrieked against his creed—*[41]

What a graphic expression of unfulfilled expectations. Tennyson's verse shows how nonplussed many Victorians were by nature's quandaries and, more important, how those quandaries shook their religious faith. The main theme of Tennyson's poem is the struggle to believe in light of natural evil. Creation was not viewed as groaning under the weight of sin but rather was expected to be harmonious and perfect. So idealistic were the Victorians' expectations that even unsprouted seeds were a problem. It seemed that God and nature were at strife.

Darwin added his own set of difficulties to the naturalist's problem of evil. As we have seen in chapters 1 through 4, Darwin found much of biology to be at odds with his idealistic expectations. Everything from the way species were classified to their habitats and behaviors seemed arbitrary and capricious. It was, he often concluded, "utterly inexplicable on the theory of creation." Nature seemed to lack precision and economy in design and was not befitting of the Creator. There were birds that laid a bounty of eggs only to have many rot, and there was the "strange and odious instinct" of the young cuckoo bird that ejected its siblings from the nest. Bees destroyed themselves with their sting and produced drones in vast numbers for one single act, only then to be slaughtered by their sisters.[42] There must have been some other explanation for all this confusion; Darwin could hardly imagine that his good God was behind "the clumsy, wasteful, blundering, low, and horribly cruel works of nature."[43]

Darwin's Solution for Natural Evil

From Aristotle to Tennyson, thinkers have always been concerned with the problem of natural evil. One solution was simply to deny the evil and, after Paley, focus on the utilitarian aspects of nature. But Darwin saw how poorly this smiling face of nature reflected reality. "I own that I cannot see as plainly as others do," Darwin wrote to Asa Gray in 1860, "and as I should wish to do, evidence of design and beneficence on all sides of us. There seems to me too much misery in the world. I cannot persuade myself that a beneficent and omnipotent God would have designedly created the [parasitic wasp] with the express intention of their feeding within the living bodies of caterpillars, or the cat should play with mice."[44]

From the Cambridge Platonists to David Hume, the idea that the world should not be seen as God's handiwork had been building for centuries.

Better for the world to be the result of a mechanistic process—God's secondary causes for the reverent, the universe's natural laws for the skeptic. Darwin summarized the notion in his letter to Gray: "I am inclined to look at everything as resulting from designed laws, with the details, whether good or bad, left to the working out of what we may call chance."[45]

The problem then, for Darwin and others sharing his view, was to derive a mechanical solution that invoked natural laws and absolved God of responsibility for nature's iniquity. The solution must drive nature by natural laws, not God's hand. But unlike the symmetrical celestial orbits of astronomy, biology seemed to be a collection of special circumstances, full of anomalies instead of the harmony bequeathed by natural law. How could science satisfy Whewell's dare with a mechanistic explanation of species if nature was so at odds with itself? Could a natural law be found to explain natural evil?

The stage was set. British naturalist Alfred Russel Wallace produced a solution that was strikingly similar to Darwin's theory of evolution. Even Wallace's language was similar.[46] Darwin's solution was that species evolve via the process of natural selection acting on random biological variation. The idea that such variation occurs naturally was obvious—look at any population and one could see a spectrum of traits. But each new generation of the population was derived *only* from those in the previous generation that survived long enough to reproduce. There was a natural selection process filtering out those traits, or combination of traits, not fit to survive.

The idea was both simple and profound. It was not, however, the sort of theory that could immediately produce specific predictions. Whereas Newton's laws of motion and gravity could predict the trajectory of a cannonball, Darwin's process of evolution provided an explanation for what was already known. And whereas Newton's theory was specific, Darwin's left much to the imagination. Evolution could have occurred in a variety of ways; in fact, just about everything found in biology could be explained with evolution. The exquisite design and adaptation of the species reflect evolution's efficiency. On the other hand, the waste and carnage in nature reflect evolution's limited scope—it only addresses reproduction. And the evolutionary framework leaves plenty of room for adjustments and subhypotheses to explain new findings. Though evolution does not provide specific and unambiguous scientific predictions, it does provide a specific and unambiguous paradigm. Evolutionists can investigate nature under the assumption that it arose naturally with no divine guidance. One need no longer be concerned about how or why a good God would make this bad world.

Sedgwick's Response

Wallace's paper prompted Darwin to finally publish his theory of evolution. As could be expected, those naturalists opposed to the use of natural law to explain the origin of species were critical. Adam Sedgwick's (1785–1873) initial response, in a letter to Darwin, was scathing. Though he admired parts of the theory, he found other parts "utterly false and grievously mischievous." The dispute was not so much over scientific reasoning as over metaphysics.

Sedgwick was confounded by Darwin's apparent disregard for the problem of morality, for Darwin had concealed the Creator behind secondary effects to the point of making him irrelevant. No Creator meant no one to give the moral law. The existence of morality is obvious, and the moral law must have come from a higher authority. Now Darwin was removing the very basis for morality. Take away the Creator, and you have no source for the moral law.

But Darwin was not blind to nature's moral implications, for it was nature's cruel and wasteful aspects that had influenced his thinking. He had been handed the God of rational theism and was having difficulty reconciling this deity with what he saw in nature. Darwin was very concerned with Sedgwick's link to morality, but he was more troubled by biological difficulties that did not seem to bother the mentor geologist. If rational theism implies a benevolent God who manifests himself in an idyllic world, then why was Darwin seeing so much imperfection? Darwin's gritty and chaotic world implied no such Creator.

Darwin's reconciliation resolved the metaphysical dilemma that bothered him but not Sedgwick—the problem of evil. But with one metaphysical dilemma gone, another stepped in to take its place—the one that bothered Sedgwick: the problem of morality. From where does our moral law derive? Within the spectrum of rational theism, Darwin traded dilemmas.

The rational theists had had God for their source of morality but could hardly reconcile him with the vagaries of nature. Darwin brought a shift, within the spectrum of rational theism, toward agnosticism and pantheism, and in doing so he stirred up the problem of morality. With God distanced from creation, evil could be explained as arising from natural law, but then how did morality arise?

As with Milton, Darwin's thought was influenced by his concept of God. Darwin believed that God would not have created the biological world as we find it and saw evolution as a way around this problem. And just as Milton's theodicy in many ways reflected the seventeenth-century view of God, Darwin's evolution was consistent with the Victorians' religious beliefs. But as always, there was a spectrum of such beliefs. For many

Victorians the problem of evil was critical, but for others such as Sedgwick, the problem of morality was more important. For our purposes,
what is important is that the metaphysics underlying evolution run deep.
Darwin's theory did not obviate metaphysics, it incorporated a particular metaphysic.

8

Evolution and Metaphysics

Charles Darwin is sometimes called an intellectual revolutionary, but this is an overstatement. Darwin accelerated a movement; he did not start one. He presented what we might call the evolution theodicy, which distanced the Creator from natural evil just as Milton had distanced the Creator from moral evil. The metaphysical ideas that influenced Darwin trace back centuries earlier.

Darwin's worldview was not revolutionary, but Darwin did find a way to describe his worldview using scientific terms. The idea that God must be aloof—separated from creation—now became more respectable. With evolution, science's stamp of approval gave further credence to the idea of a distant God.

Milton's theodicy did more than explain how God could allow for moral evil. It also affirmed views of God and humanity that throughout history have been popular with religious thinkers. God, on the one hand, is seen as all-good but not necessarily as all-powerful, or at least he does not exercise all his power. God is virtuous, not dictatorial.

In one version or another this view has been espoused by a range of influential philosophers and theologians. One twentieth-century proponent summarized it as follows: "Goodness is more fundamental than power. . . . There is nothing worthy of worship in power as such. . . . After all, the object of religious worship is a perfect ideal rather than a perfect power. . . . The limiting of the ideal by theistic absolutism is more irreligious than the limiting of power by theistic finitism."[1] In other words, a lesser God not only is distanced from evil but also becomes more sublime and worthy of our worship.

Humanity, on the other hand, now has the proper moral imperative. For some people, too much emphasis on God and his grace would leave humans without ethical motivation. It seems that the focus needs to be on human autonomy and responsibility to motivate believers and convert skeptics. Better to take a strong stand for morality, even if we lessen God's role, than to defend omnipotence and therefore undermine human morality. In religion, then, too much God detracts from the human prerogative.

Likewise in science, the naturalist needs the proper motivation. Some scientists have argued that a sovereign God who actively controls his creation is a roadblock to scientific inquiry, for we never know when a law is being suspended. A more passive God does not dabble in his creation, so scientific inquiry becomes plausible.

These two metaphysical ideas, divine sanction and intellectual necessity, have been important in evolutionary thought. They serve as a sort of theological justification for the evolution theodicy. The fact that evolution is dependent on metaphysical ideas for its justification is not often acknowledged. What is commonly discussed, however, are the various metaphysical conclusions suggested by evolution. In fact, in the history of science, evolution is probably the single most influential theory in areas outside of science. Of course, it is no coincidence that the metaphysical ideas that have arisen as a consequence of the acceptance of evolution follow from the metaphysical ideas that originally influenced the development of evolution, but the debt to those older ideas has long since been forgotten. Our new metaphysical "truths" are mistakenly viewed as the implications of evolution rather than the foundations of evolution.

This chapter gives a brief survey of divine sanction and intellectual necessity in evolutionary thought and how the acceptance of evolution has influenced our current metaphysics.

Evolution's Divine Sanction

The view that God should work according to natural laws rather than direct providence has been attractive to many naturalists. In the nineteenth century uniformitarianism and evolution were found to be theologically superior because God was thought to be thus properly juxtaposed to creation. The Creator, so the reasoning went, was all the greater for not dabbling in the details of nature. Geologists James Hutton and John Playfair contended that uniformitarianism reveals God's wisdom and therefore is far more conducive to reverent contemplation than the brute supernatural intervention of the Mosaic account.[2] Likewise, Charles Lyell thought it more worthy of God to have designed interdependency to ensure balance and uniformity.[3] In *Vestiges*, Robert Chambers wrote, "How can we suppose an immediate exertion of this creative power at one time to produce the zoophytes, another time to add a few marine mollusks, another to bring in one or two crustacea, again to crustaceous fishes, again perfect fishes, and so on to the end. This would surely be to take a very mean view of the Creative Power."[4] Divine providence could engage in the noble activity of impressing laws on matter but not grovel in the muck of nature.

Likewise for evolution's cofounder, Alfred Wallace, the universe was self-regulating according to its general laws and in no need of continual supervision and rearrangement of details. "As a matter of feeling and religion," concluded Wallace, "I hold this to be a far higher conception of the Creator of the Universe than that which may be called the 'continual interference hypothesis.'"

This approach easily strayed at times into the ancient Gnostic camp, where the physical and spiritual worlds are opposed. The physical realm is evil and the spiritual realm is good. God's work has no relation to his Word, for the creation is beneath the dignity of the Lord. The Reverend William Conybeare claimed, "The Bible is exclusively the history of the dealings of God towards men."[5] His point was that the Bible is spiritual and should not be used to infer the history of creation. The Reverend Baden Powell insisted that physical and moral problems have completely separate foundations and should have nothing to do with one another. God's works and God's Word are separate, and moral and physical phenomena are completely independent. He wrote in 1838:

> Scientific and revealed truth are of essentially different natures, and if we attempt to combine and unite them, we are attempting to unite things of a kind which cannot be consolidated, and shall infallibly injure both. In a word, in physical science we must keep strictly to physical induction and demonstration; in religious inquiry, to moral proof, but never

confound the two together. When we follow observation and inductive reasoning, our inquiries lead us to science. When we obey the authority of the Divine Word, we are not led to science but to faith. The mistake consists in confounding these two distinct objects together; and imagining that we are pursuing science when we introduce the authority of revelation. They cannot be combined without losing the distinctive character of both.[6]

Religion and science are to be kept separate. God is retained to supply the former, but it would never do to consider him in the latter. Powell resolves the problems of morality and evil by having God appear and disappear as needed. The Creator is used to explain morality but is disconnected from the physical world.

Darwin, for his part, was keen to the implications of this new Gnosticism. If God is not intimately involved in the world, then is he involved at all? In a letter Darwin challenged his American friend Asa Gray to think this through:

I see a bird which I want for food, take my gun and kill it, I do this designedly. An innocent and good man stands under a tree and is killed by a flash of lightning. Do you believe (and I really should like to hear) that God designedly killed this man? . . . If you believe so, do you believe that when a swallow snaps up a gnat that God designed that that particular swallow should snap up that particular gnat at that particular instant? I believe that the man and the gnat are in the same predicament. If the death of neither man nor gnat are designed, I see no good reason to believe that their *first* birth or production should be necessarily designed.[7]

Darwin may have been more skeptic than believer, but he knew very well how to craft a religious argument. The Scriptures proclaim that God is free to create calamity, yet that his providence extends even to birds.[8] But in the Victorian world, Darwin could question this with little justification required. It was reasonable for Darwin to argue that God would not be personally involved in the swallow's attack on the gnat—not because of any finding of modern science but because of the persistence of Gnosticism into modern times. And given such a premise, it was then reasonable to conclude that God is altogether removed from the world. Evolution is the right conclusion given a Gnostic starting point. God and matter don't mix, so life wasn't created.

Darwin also used Gnostic ideas to defend his theory against the problem of complexity. We saw in chapter 4 that Darwin defended his theory against complex organs by shifting the burden of proof to the skeptic. In addition to this, he pointed out that while it is tempting to see God as the master engineer who crafted complex organs such as the eye, this would make God too much like human beings. Darwin agreed that the perfection of the eye reminds us of the telescope, which resulted from the highest workings of human intellect. Is it not right to conclude that the eye is also the product of a great intellect? This may seem the obvious answer, but Darwin warned against it, for we should not "assume that the Creator works by intellectual powers like those of man."[9] Better to imagine the eye as the result of natural selection's perfecting powers than to have God too much involved in the world. The Victorians could not believe that the boy Jesus actually labored in his earthly father's carpentry shop. Likewise, it was reasonable for Darwin to argue that complex organs were not likely shaped by God because that would mean he works as humans do.

These Gnostic tendencies remain with us today. Evolutionist Stephen Jay Gould, for example, admiringly recounts the Darwin-Gray correspondences culminating in the above quote. The problem, according to Gould, is not that Darwin had woven a religious argument into his supposedly scientific theory but that his position can be depressing. Gould explains how we are supposed to understand this new Gnosticism, and he has invented an acronym for his principle: NOMA, or "non-overlapping magisteria." "I do not see," Gould writes, "how science and religion could be unified, or even synthesized, under any common scheme of explanation or analysis."[10]

Gould is not the only contemporary evolutionist with Gnostic sympathies. Niles Eldredge takes the position that "religion and science are two utterly different domains of human experience,"[11] and Bruce Alberts, writing for the National Academy of Sciences, says, "Scientists, like many others, are touched with awe at the order and complexity of nature. Indeed, many scientists are deeply religious. But science and religion occupy two separate realms of human experience. Demanding that they be combined detracts from the glory of each."[12]

These are just a few among many examples of modern Gnosticism within evolutionary thought. Where did Alberts learn that combining science and religion detracts from the glory of each? Certainly not from a scientific experiment. God is assumed to be disjoint from creation so that any attempt to force-fit them is bound to be awkward. Or again, how is it that God could create the universe but have nothing to do with science? The answer of course is that God did not create the world, at least not directly—the world evolved.[13]

The historian's assessment of Gnosticism could just as easily apply to evolution: "The cardinal feature of gnostic thought is the radical dualism that governs the relation of God and world. . . . The deity is absolutely transmundane, its nature alien to that of the universe which it neither

149

created nor governs and to which it is the complete antithesis. . . . The world is the work of lowly powers.[14]

The Gnostic's hope in "lowly powers" was fulfilled in evolution's natural selection. The acceptance of evolution, in turn, reinforced Gnosticism in modern thought. Darwin gave form to the Gnostic's vision, but that brought with it a movement toward Gnosticism.

The influence of Gnostic thought today is not often acknowledged or understood. It is, according to Harold Bloom, the most common thread of religious thought in America. He calls it the American Religion and finds it "pervasive and overwhelming, however it is masked, and even our secularists, indeed even our professed atheists, are more Gnostic than humanist in their ultimate presuppositions."[15] It is perhaps one of the great ironies in religious thought that one can profess to be an agnostic, skeptic, or even atheist regarding belief *in* God yet still hold strong opinions *about* God. Evolution may breed skepticism, but its adherents have continued to make religious proclamations. And those proclamations are really no different from those made by Darwin and his fellow Victorians.

Evolution's Intellectual Necessity

If God is not actively controlling creation, then it must be ruled by fixed natural laws, and this, evolutionists argue, makes scientific inquiry possible. Darwin's friend Lyell argued that uniformitarianism is required for science to advance properly, and Baden Powell agreed. In 1855 Powell wrote that not only geology but all science depends on the principles of uniformitarianism.[16] Evolutionists also found this to be the case. Though Darwin confidant J. D. Hooker found special creation and evolution at an empirical standoff, neither theory with a clear advantage, he opted for the latter for its "great organizing potential." It was not that evolutionary theories were "the truest," he wrote to William H. Harvey in 1859, "but . . . they do give you *room to reason* and reflect at present, and hopes for the future, whereas the old stick-in-the-mud doctrines . . . are all used up. They are so many stops to further inquiry; if they are admitted as truths, why there is an end of the whole matter, and it is no use hoping ever to get any *rational explanation* of origin or dispersion of species—so I hate them."[17]

For Darwin, not only did homologies argue against a Creator, but any attempt to find a divine explanation for homologies would be unscientific anyway:

Nothing can be more hopeless then to attempt to explain this similarity of pattern in members of the same class, by utility or by the doctrine of final causes. The hopelessness of the attempt has been expressly admit-

150

ted by Owen in his most interesting work on the "Nature of Limbs." On the ordinary view of the independent creation of each being, we can only say that so it is,—that it has pleased the Creator to construct all the animals and plants in each great class on a uniform plan; but this is not a scientific explanation.[18]

Here Darwin extrapolated from his metaphysical argument to arrive at the ultimate proof against creation. His main point, that nature fails to reveal a divine hand, was now protected against counterarguments, because such arguments would be unscientific—though he had repeatedly used metaphysical arguments against creation to prop up evolution.

Darwin correctly observed that creation and its supporting arguments hinge on one's concept of God, but he conveniently forgot that arguments *against* creation equally hinge on one's concept of God. He found it fair to argue against creation but not for it. Thus evolution is the correct scientific conclusion. In fact, what good science requires is a naturalistic explanation, regardless of what particular explanation is used. He wrote in 1863:

Whether the naturalist believes in the views given by Lamarck, by Geoffroy St. Hilaire, by the author of the "Vestiges," by Mr. Wallace or by myself, signifies extremely little in comparison with the admission that species have descended from other species, and have not been created immutable: for he who admits this as a great truth has a wide field open to him for further inquiry.[19]

Thomas H. Huxley saw Darwinism as a powerful instrument of research. Follow it out, he wrote, and it will lead us somewhere.[20] Today this notion persists. For Eldredge, the key responsibility of science—to predict—becomes impossible if belief in a capricious Creator is entertained:

But the Creator obviously could have fashioned each species in any way imaginable. There is no basis for us to make predictions about what we should find when we study animals and plants if we accept the basic creationist position. . . . The creator could have fashioned each organ system or physiological process (such as digestion) in whatever fashion the Creator pleased.[21]

This idea is also important for Paul Moody.

Most modern biologists do not find this explanation [that God created the species] satisfying. For one thing, it is really not an explanation at all; it amounts to saying, "Things are this way because they are this way." Furthermore, it removes the subject from scientific inquiry. One can do no more than speculate as to why the Creator chose to follow one pattern in creating diverse animals rather than to use differing patterns.[22]

Tim Berra warns that we must not be led astray by the apparent design in biological systems, for it "is not the sudden brainstorm of a creator, but an expression of the operation of impersonal natural laws, of water seeking its level. An appeal to a supernatural explanation is unscientific and unnecessary—and certain to stifle intellectual curiosity and leave important questions unasked and unanswered."[23] In fact, "creationism has no explanatory powers, no application for future investigation, no way to advance knowledge, no way to lead to new discoveries. As far as science is concerned, creationism is a sterile concept."[24]

Evolutionists have always used negative theology to argue against divine creation, but here evolutionists are not so much saying creation is wrong as that it is improper. Evolution is intellectually necessary because divine creation cannot be investigated and analyzed. For Isaac Newton and many other scientists the idea that God created the world has been a stimulus to scientific inquiry, but for today's evolutionists it is anathema.

This idea that divine creation is intellectually untenable sometimes takes the form of the "God of the gaps" defense of evolution: an appeal to supernatural mechanisms is really just using God to fill the gaps in our understanding of nature. Such an appeal, warns the evolutionist, won't work because eventually scientists will find naturalistic explanations for any gap. If God is used to explain what we don't know, he will be reduced as our knowledge grows. Therefore the "God hypothesis" should be ruled out from the very beginning.

Inherent in this thinking is a rather uncritical assessment of things scientific. As we saw in chapter 7, Darwin's contemporary William Whewell allowed that scientific theories, when "clearly proved," must be accepted. But how is a scientific theory to be clearly proved? What Whewell left undefined has remained an open question in the philosophy of science. Cardinal John Newman was a bit more cautious than Whewell when he warned of chasing scientific "phantoms"—theories that today seem right but tomorrow will be discounted.[25] But with the acceptance of evolution, the idea that science ought always to be tentative in its conclusions has given way to unreserved optimism. Evolution, we are told, is a fact, and this unguarded confidence has simply bolstered the God-of-the-gaps defense. We used to believe God must have created life, but now we know that natural mechanisms are sufficient. This conclusion hinges on the success of evolution, which in turn hinges on its concept of God. A God who must be distanced from the world and its evils was assumed, so now we conclude with a God who must not stand in the way of naturalistic explanation. The original assumption feeds right through and becomes the final conclusion. The God-of-the-gaps defense arises not from the assumption that there is no God but from the assumption that there is a particular type of God.

The result is that God is ruled out in scientific investigations. As evolutionists Maitland Edey and Donald Johanson write: "What God did is a matter for faith and not for scientific inquiry. The two fields are separate. If our scientific inquiry should lead eventually to God . . . that will be the time to stop science."[26]

But how could their inquiries possibly lead to God if they make the assumption up front that "what God did is a matter . . . not for scientific inquiry"? If one is searching only for mechanistic solutions, then that is what one will find.

We can always contrive naturalistic explanations if we try hard enough. The theory of evolution is an outstanding example. We are told that life and its enormous complexity must have arisen spontaneously, even though we don't know quite how it happened. We may not have the answers now, but given enough time we will be able to provide a mechanistic story. To invoke God is to tamper with our intellectual necessity.

Two powerful metaphysical doctrines have been used to support evolution: divine sanction with its Gnostic overtones, and intellectual necessity. Evolution is seen to serve both the Creator and the creature—it is simultaneously more befitting of God and more empowering of humankind. God is properly distanced from creation, and humans are properly motivated to explore creation.

But although God may seem to have gained stature, the focus is now shifted away from him. The argument that God becomes all the more worthy of our reverence easily gives way to his loss of relevance.

Materialism

Theodicies for both natural and moral evil push God into the background. Taken to the extreme, this leads to atheism and materialism, with the universe as nothing but matter and motion. Huxley strayed in this direction with his notion of animals and people as biological robots, or as he put it, "conscious automata." In such a world there is no authority that supplies our sense of morality, and therefore judgments regarding evil arise only from our personal feelings. Oxford zoologist Richard Dawkins summarizes the view: "The universe we observe has precisely the properties we should expect if there is at bottom no design, no purpose, *no evil and no good*, nothing but pointless indifference."[27]

The strength of materialism is that it obviates the problem of evil altogether. God need not be reconciled with evil, because neither exists. Therefore the problem of evil is no problem at all.

But it is difficult for the materialist to remain entirely consistent, for who is ready to say the worst examples of evil are not, after all, really evil? The mass murderer and the healer are just acting out their respective impulses with no ultimate moral implications attached. And of course since there is no evil, the materialist must, ironically, not use the problem of evil to justify atheism. The problem of evil presupposes the existence of an objective evil—the very thing the materialist seems to deny. The argument that led to materialism is exhausted just when it is needed most. In other words, the problem of evil is generated only by the prior claim that evil exists. One cannot then conclude, with Dawkins, that there is "no evil and no good" in the universe.

If the materialist must not be motivated by the problem of evil, where does this leave Darwin, who saw evolution as the explanation to the evil he saw in nature? Although Darwin's gritty world cried out for justice, his evolution would grant none, for it would call back that there was no injustice in the first place. Evolution, if anything, would have us admire the wasp's parasitic adaptability. And so the problem of evil is solved because there is no evil, but this leaves unresolved the feelings of sensitive people like Darwin. Such feelings of injustice must be denied, for they aren't real. In *Origin* Darwin tried to explain why nature has evil but not why we perceive it *as* evil. Hence Sedgwick's complaint that Darwin had ignored the link between nature and morality.

But all was not lost. It may have been too bold for Darwin to suggest this initially, but as evolution gained acceptance it became possible to ascribe morality to evolution itself. Did not the moral sense have utility in the evolution of civilization? As the monkey fears the snake, Darwin wrote in his autobiography, so the child believes in God.[28] Though Darwin's wife excised this passage, the notion that faith and morality emerged from the evolutionary process persisted. Evolution gave rise not only to all life but to the spiritual sense and moral law as well. What Kant had found to be sufficient to prove God—the moral law within—Darwinism subsumed with ease. Morality does not transcend the material world but is a product of natural selection.[29] Indeed religious belief itself is nothing more than a byproduct of evolutionary history, a phenomenon rooted in our genes. "The human mind evolved," writes evolutionist Edward Wilson, "to believe in the gods."[30] And by relentless testing, Wilson predicts, science will uncover the real source of those "moral and religious sentiments."[31]

Science, it seems, is our new source of truth, and it reveals that religion and morality were created by evolution. What is not acknowledged, however, is that the main arguments for evolution appeal to this sense of morality—they are not scientific. Remove these, and weak arguments about evolution's plausibility are all that remain. By not acknowledging the foundational moral arguments, evolutionists have mistakenly concluded that evolution is good science. Their strong confidence in Darwin's theory has spurred them to make all sorts of metaphysical extrapolations.

Evolution is now found to be capable of creating just about anything. We might say that evolution is a closed metaphysical system. It not only supplies its own creation story but also supplies its own source of morality. Both were products of the evolutionary process. Furthermore, having rejected divine creation and its Creator, evolution even becomes its own authority. This story is true for those who believe it, but it cannot be demonstrated by strictly scientific argument, for it requires metaphysical premises.

To summarize, then: The twin pillars of evolution and materialism seem to resolve the problem of evil, but their foundations go deep into metaphysics. The problem of natural evil had become increasingly acute as modern naturalists uncovered the seemingly irrational inner workings of nature. As we saw earlier, Darwin's theory of evolution was a response to this.

We often hear that evolution is an objective, scientific theory. The theory may affect our metaphysics by virtue of its powerful scientific basis, but its formulation stands in a metaphysical vacuum. Though it generates its own metaphysics as output, it takes none as input.

But this is a great myth of our time. Evolution is not a story of a bold scientific stroke that has been beautifully borne out by the advancement of science, against metaphysical resistance. It is nearly the exact opposite. It is not that evolution is utterly unscientific or that it completely lacks evidence. Evolution can be properly formulated as a scientific theory with plenty of supporting evidence, but as such it is unremarkable. Evolution's supporting evidence is outrun by the counterevidence. Both nineteenth- and twentieth-century science provided more than enough challenges to put evolution's validity in doubt, but the nineteenth century's metaphysical trends have continued through and beyond the twentieth century. Evolution's compelling arguments, and the reason for its stunning success, come not from its scientific support but from indirect arguments against creation.

Phillip Johnson argues that in order to be a science rather than a branch of philosophy, evolutionary biology must pose scientific questions rather than question the motives of God, but the point has been lost on evolutionists. Indeed in his critique of Johnson's *Darwin on Trial*, David Hull showed how the nineteenth-century arguments against creation persist today:

> What kind of God can one infer from the sort of phenomena epitomized by the species on Darwin's Galápagos Islands? The evolutionary process

is rife with happenstance, contingency, incredible waste, death, pain and horror. Millions of sperm and ova are produced that never unite to form a zygote. Of the millions of zygotes that are produced, only a few ever reach maturity. On current estimates, 95 per cent of the DNA that an organism contains has no function. Certain organic systems are marvels of engineering; others are little more than contraptions. When the eggs that cuckoos lay in the nests of other birds hatch, the cuckoo chick proceeds to push the eggs of its foster parents out of the nest. The queens of a particular species of parasitic ant have only one remarkable adaptation, a serrated appendage which they use to saw off the head of the host queen. To quote Darwin, "I cannot persuade myself that a beneficent and omnipotent God would have designedly created the *Ichneumonidae* with the express intention of their feeding within the living bodies of caterpillars."

Whatever the God implied by evolutionary theory and the data of natural history may be like, He is not the Protestant God of waste not, want not. He is also not a loving God who cares about His productions. He is not even the awful God portrayed in the book of Job. The God of the Galápagos is careless, wasteful, indifferent, almost diabolical. He is certainly not the sort of God to whom anyone would be inclined to pray.[32]

Hull's bold use of evolution's negative theology shows how ingrained Darwin's arguments have become. The God that Darwin began with has now become the de facto standard.

The Ultimate Defense of Evolution

Another defense that evolutionists sometimes use against critics such as Johnson is that despite all its problems, evolution remains the best explanation of the scientific evidence. How important can criticism of evolution be if the critic doesn't also supply a better theory? After all, the species came about somehow. If not by evolution, then how? Until a better explanation comes around, we can hardly drop evolution just because there are some problems with the theory. The right response, say evolutionists, is not to drop the theory but to address the problems. That evolution is a fact does not mean all the details have been worked out. Criticism suggests paths for future research; it does not mean evolution is wrong.

This might be called the *no-alternative* defense of evolution. Not only must the critic show evolution to be flawed, he or she must also solve the problem. Falsifying evolution is not good enough, for evolution is presumed true until a better solution is provided.

But this is not the way science is supposed to work. In this defense evolution has been given a special status that has no place in science. Unlike defendants in our legal system, who enjoy the presumption of innocence until guilt is proved, theories are not granted a priori any special status. They are certainly not presumed to be true simply because an alternative has not yet been proposed. Nor are they to be presumed true because they have not been shown to be false. Such presumptions are called *informal fallacies*,[33] and they are practically the antithesis of scientific thinking.

The no-alternative defense gives evolution a special status not normally accorded to scientific theories. In fact, the claim that evolution is the best explanation available is itself a nonscientific statement. Evolutionists have repeatedly argued that their theory works far better than the notion of divine creation, but in so doing they have made substantial assumptions about the nature of God. Their negative theological arguments are not scientific. Evolution is the best explanation of the scientific data only if one adopts a particular metaphysical view.

Metaphysics can complicate things terribly, but such complications do not typically arise in science. Science has the good fortune to be free of metaphysical quandaries not because, as some have supposed, it is free of metaphysics altogether, but because the metaphysics are held to a constant. In most cases, competing scientific theories share the same assumptions about reality—for example, that natural laws are uniform over space and time. The metaphysics are not what is in question. One can argue about how to interpret an experimental result without ever getting stuck in metaphysical puzzles, but evolution is different in this regard. Here scientists have strayed into an area where the competing ideas do not share common presuppositions about the world. Consequently the competing ideas cannot be compared on a purely scientific basis. The history of evolutionary thought bears this out clearly. The plethora of metaphysical language in the evolution literature shows how the debate is about more than just science.

In the eighteenth century, philosopher David Hume criticized natural theology, the use of science to prove God. According to Hume, natural theologians were going well beyond the scientific data when they claimed to have found evidence for their God. If ten ounces are raised on a scale, then the counterbalancing weight must be at least ten ounces, but that is all one can conclude. One cannot conclude that the counterbalancing weight exceeds one hundred ounces. With this analogy Hume exposed the natural human tendency to infer unwarranted conclusions from an experiment. Now, more than two centuries later, Hume's skepticism rings true, this time as a warning against evolution's grandiose claims. Evolution claims to prove that there is no divine hand evident in nature. God, if he

157

exists, need not be considered when we investigate nature, but one cannot find evidence against the divine without first assuming something about the divine. Evolutionists have superimposed their idea of God over creation and found that they do not match up very well, but the mismatch depends every bit as much on the theology as on the science.

The evolutionist's refrain that evolution is still the best explanation leads to a logical fallacy, but more important, it reveals the nonscientific nature of evolution. Criticism of evolution should not be required to supply a better scientific explanation, because evolution was never scientific to begin with. Evolution is an organizing idea that inherently relies on ultimate truth claims—claims that are outside of science. Evolution draws on several scientific disciplines, but evolution itself is not scientific. Thus it is not a matter of finding a better scientific explanation before evolution is dropped from science; rather, it is a matter of understanding the boundaries of science. When assumptions about God are made before the science begins, the result is not science, no matter how much science follows. When limited to scientific findings, the evidence for evolution is easily countered. And when the metaphysics are brought in for support, evolution relies on a particular and rather sentimentalized version of God. The evolution theodicy is a combination of questionable science and narrow metaphysics.

Needed: A Tap on the Shoulder

Many wonder why evolutionists make such high claims of success while the theory incurs scientific difficulties that would do away with most theories. The answer is that evolutionists find their confidence not in positive arguments for evolution but in negative arguments against the modern idea of creation. When evolutionists claim that a particular scientific observation proves their theory, they are not committing the fallacy of affirming the consequent of the premise they wish to prove; rather, they are denying the consequent of the premise they wish to disprove. Evolution is proved not because it is verified but by the process of elimination. As Ernst Mayr wrote, it must be admitted that Darwinism has achieved acceptance less by irrefutable proofs in its favor and more by the default of opposing theories.[34]

As we saw, Darwin's arguments are against the doctrine of creation as much as for evolution. Certainly this is the form of the very compelling arguments that he and his disciples advanced—the arguments that allow modern evolutionists to refer to evolution as a fact. Transmutation and genealogical connection are facts not from science but from arguments against special creation. Darwin wrote in *Descent of Man* that he had two distinct objectives in advocating his theory: first to show that species had not been separately created, and second to show that natural selection was the chief agent of change. It was the former that was primary for Darwin; he could forgo natural selection. "If I have erred" by exaggerating natural selection, Darwin explained, "I have at least, as I hope, done good service in aiding to overthrow the dogma of separate creations."[35]

One can try to cast evolution in a scientific mold, but ultimately it is wed to its metaphysical presuppositions about the nature of God. Alfred North Whitehead observed that when critiquing a tradition of thought, one should not focus most on those arguments its exponents put forth but on the fundamental assumptions that seem obvious and are unconsciously presupposed.[36] We see consistent and unabashed references to the God of rational theism interwoven in the scientific discussions of Darwin and his modern disciples. In the name of science we are told just what God is and is not capable of doing, for creation is defined according to rational theism's pacified God who could not possibly, for example, have created all those species of beetle. Modern pundits of evolution issue their theological dictums without apology, apparently oblivious that they have strayed into foreign territory where their science will do them no good.

Evolution is deeply wedded to its metaphysical presuppositions. A particular doctrine of God is a prerequisite for evolution's success. It is a theological view that preceded evolution historically and became the metaphysical landscape on which the theory was constructed.

The fact that evolution's acceptance hinges on a theological position would, for many, be enough to expel it from science. But evolution's reliance on metaphysics is not its worst failing. Evolution's real problem is not its metaphysics but its denial of its metaphysics. Philosophers warn that sometimes those who deny being under any metaphysical influence are the most influenced of all. This warning applies to evolution. Ludwig Wittgenstein said that philosophy is a tap on the shoulder, reminding the specialist what lies in the background of any area of study. Evolution is in need of such a reminder.

Evolutionists routinely employ metaphysical premises, but they just as routinely claim the higher ground of pure science. The metaphysical premises are unspoken and wrapped up in technical language. The result is a scientific-sounding theory that preaches its own ultimate truths. What is apparently nothing more than empirical observations and logical deductions somehow is able to arrive at the true nature of reality. It tells us that God, if there is one, is benign. He may have created the world, but he has shown little interest since then. The formation of life itself has been left to its own devices—a random, unguided process—

while God hides in the recesses of the universe. We can understand evolution only when we see that the God evolution found is the same God that evolution started with. Its evidences and arguments led to that God only because they presupposed that God.

Evolution is no more objective than any of the other theodicies that preceded it. Evolution provides a theological solution to a theological problem, and the science is sandwiched somewhere in between. But the theological premises are denied, so the theological result is seen as coming from science, and science inappropriately attains the status of truth-giver. Philosophy and science have always been influenced by theology. This is especially true for evolution. The difference is that evolution denies the influence.

9

Blind Presuppositionalism

The theory of evolution is one of the great examples of humanity's quest for objective knowledge. It proposes to explain the world without presupposing anything about the world. But is this possible? Seventeenth-century philosopher René Descartes attempted to do this sort of thing. His first principle was to accept as true only that which he knew for certain to be true. From there he used proofs according to the laws of logic. Descartes was also a mathematician, and he was impressed with mathematical proofs that derived new axioms based only on known premises. In similar fashion, Descartes tried to find objective truths about the world that relied only on what he thought were provable premises, such as self-existence. "*Cogito ergo sum*"—I think, therefore I am—was his famous starting point. From there Descartes proved the existence of God and truth without first assuming either one.

Descartes's approach was found to be faulty, but his quest for objective knowledge was taken up by many later thinkers. In the eighteenth century, for example, David Hume used another dubious set of proofs to argue that miracles are impossible. Descartes's theism contrasts with Hume's skepticism, but for our purposes the similarity in their approaches is more important. Both Descartes and Hume believed that logical argument could produce ultimate truths, and not surprisingly both found truths that were remarkably similar to their own personal beliefs. Descartes the theist found God, and Hume the skeptic found materialism.

It is important to understand the presuppositions that lie beneath an argument. Descartes and Hume incorporated certain assumptions about the world without clearly identifying them, and when those assumptions are brought to light they illuminate the rest of the argument. Likewise, the theory of evolution hinges on some fundamental nonscientific understandings that are not commonly acknowledged. There is, after all, little to be gained from advertising one's presuppositions. Such an advertisement would serve as one big caveat undermining all subsequent conclusions. Instead, it is natural to forget the presuppositions and dwell on the good work that follows. Presuppositions are very important, yet they are not often revealed.

Most evolutionists today are not particularly aware of evolution's theodicy. The vast majority of work in evolutionary studies is concerned with detailed questions about how evolution works, not the overarching arguments for why evolution is supposed to be true.

Whether objective knowledge, independent of presupposition, is even possible is a matter of philosophical debate. The theory of evolution would be an interesting case study, not as a model of how objective knowledge might be arrived at, but as a model of how subtle the use of presupposition can be. To understand evolution one must understand its metaphysical influences, but ever since Darwin, evolution has been advertised as an objective conclusion—a neutral path to knowledge. According to its adherents, the theory of evolution is not beholden to any religious or otherwise metaphysical assumptions.

But in fact the theory of evolution relies on the belief that God never would have created the world as we find it. We saw in chapters 2 through 5 that Darwin and his disciples have openly used this negative theological argument. How is it that they can simultaneously claim evolution is just a scientific theory? Furthermore, how is it that evolution prevails as the officially accepted scientific explanation for how the world came to be? The answers to these questions may well be complex, but one likely reason is simply that the nineteenth century's concept of God has continued to be popular. The fact that evolution hinges on a particular concept of God is less obvious when that concept coincides with one's personal belief about God. Rather than being nonplussed by evolution's metaphysical arguments, the twentieth century found them to be persuasive.

Many theists have attempted to integrate evolution with their faith. This tradition has been called *theistic evolution*, and it has many variations—so many that a single category seems insufficient. At one end of the spectrum, some dilute evolution to the point that they can hardly be called evolutionists. And at the other end there are those who dilute theism to the point that they can hardly be called theists. In between these two extremes there are a variety of intermediate positions.

It seems that very different worldviews can claim to incorporate evolution, to the point that historians have found it a challenge to analyze or even define evolution.[1] This has caused difficulty for evolution's critics, for how can they take aim if there is no target? This confusion can be cleared up when we understand the importance of evolution's metaphysical presuppositions. Darwin presented his theory with a long, meandering argument that was not always easy to decipher. Since Darwin, both proponents and critics have not often seen how evolution serves as a theodicy. The result has been a somewhat confusing public debate where competing views are often talking past one another. This chapter examines the various responses to evolution and how they can be understood in terms of their treatment of evolution's presuppositions.

Charles Hodge

Fifteen years after Darwin published his *The Origin of Species*, Princeton theologian Charles Hodge expressed his opinion that Darwinism was effectively antireligious in his *What Is Darwinism?*[2] Hodge questioned not Darwin's scientific findings but his interpretations of them. One of Hodge's key points was that Darwin resorted to his naturalistic explanation unnecessarily. It seemed to Hodge that Darwin had a naturalistic agenda.

For example, Darwin wrestled with the problem of how the unguided forces he was proposing could ever produce complex structures such as the eye. The notion seemed, Darwin confessed, "absurd in the highest possible degree." But Darwin went on to solve the problem with a thought experiment. He argued that we only need imagine a series of intermediates that lead up to the eye, each conferring some sort of advantage to the creature. The argument hardly seems compelling, but Darwin's point was that we need not find a *likely* series of intermediates; a merely *conceivable* series of intermediates will do. If it could be shown that a certain organ could not possibly have evolved by such a series of intermediates, then evolution would be falsified, but Darwin could find no such organ.[3] Of course this is no surprise, for one can always conjure up a hypothetical series of intermediates, especially when one is not required to demonstrate the process experimentally. In fact, falsifying evolution by Darwin's criterion is tantamount to proving a universal negative. Evolution became the default solution, true not by virtue of verification but by its nonfalsifiability.

For Hodge, Darwin's mental gymnastics revealed an underlying intention to find a naturalistic explanation at any cost. Why did Darwin not simply admit an underlying teleology beneath his law of natural selection? If God made complex structures, it could have been by the process of evolution. "But instead of referring them to the purpose of God," wrote Hodge, Darwin "laboriously endeavors to prove that they may be accounted for without any design or purpose whatever."[4] It seemed clear to Hodge that Darwinism *was* atheism.

Looking back with a century of hindsight, historians are mixed in their assessment of Hodge's criticism. Some historians say he was prescient—one of the first to see where evolution was really headed. Some of Hodge's contemporaries hoped that the theory of evolution would develop into a form that could accommodate a Creator. But in later years, as evolution gained momentum, it became increasingly clear that the orthodox version of evolution would be exclusively naturalistic. And it is this version that is presented in today's textbooks and classrooms. Hodge's warning is today's reality.

Other historians say Hodge missed the point—evolution leaves plenty of room for theism. If nothing else, this is proved by the many theists who have incorporated evolution into their beliefs. Darwin himself ended his book with an argument from divine sanction. Evolution, said Darwin, allows for a greater view of life and the Creator who originated the process. And while many interpreted this as Darwin's merely softening the blow, others continued in this thinking, retaining God and integrating theology with evolution.

Both views of Hodge are accurate to a certain degree, but neither captures the bigger picture. In developing evolution, Darwin clearly was *not* operating from an antireligious dogma. In fact, Darwin was pursuing an explanation of the relationship between the Creator and his creation. Creation did not seem befitting of the Creator, so the relationship had to be adjusted. Instead of God's creating the species, there must have been natural laws between God and his creation that did the work. The laws were blind and impersonal, and this accounted for nature's lack of harmony and symmetry.

Darwin's mental gymnastics to avoid using God in his theory, even when it came to incredibly complicated structures such as the eye, were not due to an atheist agenda, as Hodge had charged. To Hodge it seemed an easy thing to include God to account for complexity—why hadn't Darwin done so? But here Hodge did miss the point. The whole point of Darwin's theory was to separate God from the world in order to explain its inefficiencies and quandaries; he couldn't then smuggle God back into the theory to explain complexity. Rather than saying that evolution is *anti*religious, it would be more accurate to say that evolution *is* religious. It very much hinges on a particular type of God—one who would only create a world suited to our tastes.

Hodge failed to see how Darwin's concept of God, so different from his own, eventually led to the evolution theodicy. To be fair, Hodge had to do without the tremendous Darwin scholarship that has since taken place. It would have been difficult for Hodge to be aware of the thinking process that Darwin went through in developing his theory. Hodge noted that Darwin referred to "the Creator" in *Origin*, but Hodge wasn't sure quite what to make of it.

Theistic Evolution

The end result of Darwin's theory is not that there is no God but rather that God is disjoint from the material world. For some the distinction may seem trivial, for a God who is not involved in creation is no God at all. But for others the distinction leaves open the possibility of a Creator God who relies on secondary causes—the results of his divine laws of nature.

In any case, the upshot of evolution is that God, whether he exists or not, has no active role in nature. Thus evolution was a new way of viewing nature, and from this came a new research program, something that Darwin foresaw and expounded upon at length in the closing pages of *Origin*. Now free from any predilections about a divine influence in creation, human beings could research the problem and find the truth for themselves. After Darwin, this naturalistic assumption became pervasive in science—to the point that it serves as evidence *for* evolution. From Le Conte to de Beer and Ridley, evolutionists for over a century now have used this naturalistic bias in their arguments for evolution.

The naturalistic bias, as well as the antiteleological language found in arguments for evolution, is not necessarily the result of an atheistic urging. In the evolution theodicy the Creator must be disjoint from creation, but no more than this is required. For those who can accept this sort of God, theistic evolution is a possibility.

Theistic evolution is, if anything, diverse. Thinkers have tried to unite theism and evolution using just about every variety of the two domains. In most cases, however, there is a tradeoff between the two. Toward one end of the spectrum, as noted earlier, the evolution is nearly orthodox and the theism is diluted, while toward the other end the theism is orthodox and the evolution is diluted. These two camps are represented, for example, by Theodosius Dobzhansky and Hodge's successor at Princeton, B. B. Warfield, respectively.

Dobzhansky was one of the leading evolutionists of the twentieth century. He very much adhered to evolution's tenets that variation is random and unguided, that there are no final causes, and that evolution is a proven fact.[5] For example, in 1955 he wrote: "Evolution does not strive to accomplish any particular purpose or to reach any specific goal except the preservation of life itself. Evolution did not happen according to a predetermined plan."[6]

It followed for Dobzhansky that there was no purpose behind the evolution of humans. He did write about God and ultimate reality, but his thoughts were heavily influenced by his belief in evolution. He admiringly quoted Pierre Teilhard de Chardin when the latter elevated evolution to the status of a metaphysical truth:

> Is evolution a theory, a system, or a hypothesis? It is much more—it is a general postulate to which all theories, all hypotheses, all systems must henceforward bow and which they must satisfy in order to be thinkable and true. Evolution is a light which illuminates all facts, a trajectory which all lines of thought must follow—this is what evolution is.[7]

Warfield saw things quite differently. Though he described himself as a "Darwinian of the purest water," his version of Darwin's theory sharply contrasted with that of Dobzhansky. Warfield advocated a theistically directed evolutionary process that included the "constant oversight of God in the whole process, and his occasional supernatural interference for the production of *new* beginnings by an actual output of creative force, producing something *new*, i.e., something not included even *in posse* in preceding conditions."[8]

Warfield accepted and even defended this brand of evolution, but he also viewed evolution as a theory that could very well turn out to be wrong. His main point was that the evolutionary process could have been a sort of creation tool used by God, and in this sense it did not conflict with the Scriptures.

Dobzhansky and Warfield both believed in God and in evolution, but otherwise their views were so different that they can hardly be placed in the same category. One way to understand their differences is to see how they interpret evolution's theodicy—Dobzhansky embraced it, and Warfield rejected it. Dobzhansky routinely enlisted Darwin's type of metaphysical arguments against divine creation to convert scientific observations into evidences for evolution.[9] On the contrary, Warfield made it clear that his version of evolution explicitly included God's creative hand. Warfield was careful to separate the scientific version of evolution from its metaphysical version, and he rejected the metaphysical arguments for evolution, such as those presented in his day by Joseph Le Conte.[10]

Dobzhansky's and Warfield's interpretations of evolution's theodicy were opposed. This is no surprise when one considers their respective religious beliefs. Dobzhansky's God was rather abstract and impersonal and had little influence in his evolutionary studies.[11] Warfield, on the other hand, held to a high view of God's providence, so he rejected the random, unguided aspect of evolutionary theory. Dobzhansky would have rejected Warfield's view of evolution, and Warfield would have rejected Dobzhansky's view of God. Warfield did not dilute his concept of God, so his views were acceptable to those theists who had rejected evolution. Likewise, Dobzhansky did not dilute his concept of evolution, so his views were acceptable to even those evolutionists who rejected God. In other words, the two sides of the spectrum tend to diverge and merge imperceptibly into their extremes.

Dobzhansky and Warfield represent two different popular facets of theistic evolution that are opposed to each other. Toward the middle of the spectrum there are those who have attempted to maintain and reconcile orthodox views of both theism and evolution. It seems like an awkward fit, since the very idea of divine creation implies purpose rather than a random process, and any theory that forces the two ideas together is liable to draw fire from both edges of the spectrum. Theistic evolution is a tradition whose fringe thinkers may be less controversial than its centrists.

One centrist who has tried to clarify this position is biochemist Terry Gray. Gray professes belief in a sovereign God but in his scientific writings sounds very much like an orthodox evolutionist. For example, he argues that the classification of the different species is evidence for evolution. Gray does not give his reasoning, but historically evolutionists have used the classification evidence in their negative theological arguments. For example, after discussing the groupings of species, Darwin concluded that "these are strange relations on the view that each species was independently created" and that it was "utterly inexplicable on the theory of creation."[12]

For Gray, classification becomes all the more powerful evidence when different classifications are compared. Two different classifications, one based on visible features and the other based on molecular data, will usually produce similar groupings of the species under consideration. "I think that these really are," Gray writes, "independent evidences for common ancestry."[13] Again Gray does not explain how he reaches this conclusion, but evolutionists such as Ridley and Penny (see chapter 2) who have used this argument ultimately rely on metaphysical premises. Ridley, for example, writes that the similarities in different classifications "would be very puzzling" if the species were independently created.[14]

In addition to the argument from classification, Gray uses the argument from homology and imperfect design to support his view of theistic evolution. He considers vitamin C synthesis in mammals. All mammals are able to synthesize vitamin C except guinea pigs and primates, and Gray finds this to be evidence for evolution: "This in itself suggests that the ancestral mammal had the vitamin C synthesis capability, but that the ancestral guinea pig and the ancestral primate lost that ability and passed the defect to its ancestors."[15]

It is not clear how Gray completes his proof without some sort of commitment to naturalistic explanations. Nonetheless, he finds yet more evidence for evolution in the fact that a vitamin C pseudogene has been found in guinea pigs and humans. A pseudogene is a DNA sequence that resembles a gene but appears to be nonfunctional. In evolutionary lore, these are vestigial organs at the molecular level. And just as the vestigial organ argument for evolution relies on the assumption of full knowledge about the organism, so too the pseudogene argument assumes that we can be sure they are not useful. They are assumed to be the byproduct of useless, but not terribly harmful, mutations. Gray writes:

> Further analysis shows that this gene is a pseudogene, i.e., it looks like a real gene, but it is not expressed due to a mutation in the gene itself or in the region of DNA that controls the expression of that gene. Now we could argue that in God's inscrutable purpose he placed that vitamin C synthesis look-alike gene in the guinea pig or human DNA or we could admit the more obvious conclusion, that humans and primates and other mammals share a common ancestor.[16]

Here Gray makes a negative theological argument. He seems comfortable in assuming just what God would have done when it comes to designing the genotype. Gray states unequivocally that the pseudogene is a result of mutation, but this is nothing more than evolutionary speculation. More important, he then claims that God obviously would not have an inscrutable purpose for having the nonstandard gene there. For our purposes the point is not that pseudogenes do or do not have function or that God must have or must not have designed them. The point is simply that, like evolutionists, theistic evolutionists need Darwin's negative theology.

Another notable thinker who has maintained that evolution is the creative tool of God is Howard Van Till, professor emeritus of physics at Calvin College. Most people who share Gray and Van Till's view do so because they see evolution as a compelling scientific conclusion. Van Till is no exception. He makes no attempt to present a scientific argument for evolution, for he takes as his starting point that evolution is "highly credible."[17] And like Gray, Van Till is persuaded by evolution's theodicy. For example, Van Till believes that the noncoding portions of DNA reveal patterns that would not be expected if God directly created the species. Van Till writes that if those portions of DNA were created, "they would have to be considered mischievously misleading."[18]

Van Till's primary argument for evolution recalls Leibniz's theodicy (see chapter 6). Isaac Newton believed that divine providence was necessary for the maintenance and adjustment of the universe, but for Leibniz, if God intervened against his own laws, then he contradicted his own creation. Leibniz believed that the universe was optimally created by God with little need of subsequent intervention. A God who tinkers with his creation was unacceptable to Leibniz. Likewise Van Till argues that the notion that God created the universe and then subsequently set about creating species at different times is "theologically awkward,"[19] for this would mean that God withheld capabilities and then later imposed form by coercing his insufficiently equipped creation. Better for God to infuse all the capabilities at the beginning. This view easily leads to the divine sanction with its Gnostic overtones, and this is Van Till's main theme. He writes that God's transcendence over creation is more apparent if the initial creative act is followed by evolution according to natural laws rather than by a series of creative acts. In the former view creation is robust and fruitful, properties that "attest to the greatness of the Creator."[20]

In chapters 6 through 8 we saw examples of how influential the divine sanction has been. For example, Thomas Burnet in the seventeenth century, Darwin's grandfather Erasmus Darwin in the eighteenth century, and Darwin's contemporary Alfred Wallace in the nineteenth century all argued that a God who works via natural laws is all the more worthy of our reverence. Similarly Van Till sees in evolution a Creator who is all the more magnificent for creating the world not by fiat but by preplanning. As a way of emphasizing this point, Van Till prefers to call this doctrine "the optimally gifted creation" rather than "theistic evolution."

"Join with me," Van Till writes in a Gnostic-sounding rallying cry,

in celebrating the astounding giftedness of the creation as a manifestation of God's unfathomable creativity and unlimited generosity. And join

169

with me also in experiencing the Creator's lordship and transcendence over the creation, not in exceptions to the creation's giftedness, not in claims for evidence of gifts withheld, not in discontinuities, but in every gift of being that God has given to the remarkable creation of which we are an integral part.[21]

Gray and Van Till have tried to explain how Darwinian evolution could be a creative tool used by God. The views they have developed raise two main problems. First, they take evolution to be a compelling scientific finding. But we saw in chapters 2 through 5 that while there is some evidence in support of evolution, there is also a substantial body of scientific evidence that argues against evolution. From a scientific perspective, evolution is not a persuasive explanation of the origin of species. The second problem is that evolution is bolstered by nonscientific arguments about how God would not have created the species in our world. There is apparently too much inefficiency and lack of symmetry for God to have done the creating—so the arguments go. On the one hand Gray and Van Till subscribe to these arguments, yet on the other hand they believe that the design and ultimate creation of the species are actually God's handiwork. God used evolution as his tool, but he was fully in control of the tool, and the end result was according to his will. There is a contradiction here. One cannot simultaneously maintain that God would not have created the particulars of this world and that God did create the particulars of this world.

A slightly different approach to theistic evolution is advocated by biology professor Kenneth Miller. Like Gray and Van Till, Miller professes to be a Christian, and he seeks an understanding of evolution that allows God to remain in the picture. But if Van Till tends toward the Gnostic's transcendent God, Miller envisions an "at risk" deity. According to Miller, God does not control the evolutionary process, nor can he even predict just where it will lead. This should not be difficult for Christians to grasp, for, according to Miller, "Christians know that chance plays an undeniable role in history."[22] In fact, Miller considers the idea that history is not directed by God rather uncontroversial and obvious. Replay the "tape of history," Miller tells us, and "the twentieth century could easily have been very different—the next century more different still."[23] Nonetheless, Miller argues that God can use the historical process for his own ends, presumably by waiting for the right events to occur.

Miller applies this same sort of logic to natural history. Why should it be different from human history? The evolutionary history of life is full of undirected events, but God can still use it. Miller believes that the evolutionary process eventually gave "the Creator exactly what He was looking for."[24] God may not know where the world is headed, but he can

watch and wait for the right events to come along. Miller is surprised
that this view is not more intuitive for Christians. "Obviously," he writes,
"few religious people find it problematical that their own personal existence might not have been preordained by God . . . but strangely, some
of the very same people find it inconceivable that the *biological* existence
of our species could have been subject to exactly the same forces."[25]

Miller might not be so surprised if he knew a bit more about "religious
people." His ideas about what they believe are almost the polar opposite of the teachings of the Scriptures they hold dear. Scripture tells the
believer that God made her and knew her before she was born.[26] And
regarding history, God tells the believer that he declares the end from
the beginning, that what he has said he will bring to pass.[27]

Miller's apparent confusion may be due to an influence of process theology, a twentieth-century movement that has tried to answer difficult
philosophical questions about God and the world, often at the expense
of scriptural accuracy. Process theology questions such doctrines as the
Trinity, God's miracles and foreknowledge, and human sin in its attempt
to nail down metaphysical truths. Though Miller doesn't specifically use
the term, many of his ideas and sources come from process theology.[28]

Process theologians are interested in, among other things, account-
ing for modern science in their theology. Their concept of God incor-
porates scientific findings such as evolution and quantum mechanics.
Evolution suggests a God who did not directly create the world; quantum mechanics (via the Heisenberg uncertainty principle) suggests a
God who lacks full knowledge of the world and where it is going. But
such limitations, according to process theologians, can be viewed as
virtues. Certainly the problem of evil becomes easier to grapple with
given a more limited God. Also, the lowering of the Creator leads to the
raising of the creature, as God waits on the actions of human beings. The
Creator is not a cosmic tyrant, and human beings have room to grow.

Miller has elaborated at length on how we are to reconcile God and
evolution, and his ideas are very much along the lines of process theology.
First, for those who may still have doubts about the veracity of evolution,
Miller makes liberal use of the problem of evil and Darwin's theodicy to
disallow any notion that God may have created the world. Would God,
Miller asks rhetorically, "really want to take credit for the mosquito?"[29] The
answer to this and the many other quandaries of nature is obvious for Miller:
God must not have designed the world. Furthermore, undirected evolution is the obvious alternative, because God isn't allowed to control his
creation anyway. "The freedom to act and choose enjoyed by each individual in the Western religious tradition," states Miller, "requires that God
allow the future of His creation to be left open."[30] In other words, evolution must be true and it must be independent of God; otherwise "how could

the future truly be open?"[31] In Miller's view, this type of God gives us mortals the independence we need for true goodness.[32]

The point is not that Miller's system is right or wrong but that he has failed to formulate a version of theistic evolution that genuinely reconciles evolution with anything faintly resembling the traditional concept of God. We must, with Miller, accept the God of process theology who watches and waits for the world to turn his way.

John F. Haught, professor of theology at Georgetown University, also uses ideas from process theology in his formulation of a theology of evolution. Like Miller, Haught professes to be a Christian and an evolutionist. Haught makes no attempt to justify evolution, for he takes evolution to be a fact that science has uncovered. Darwin's appeal to nature's evils is, for Haught, not an unscientific liability but a challenge for anyone trying to reconcile God and evolution. Evolution is a fact, so its metaphysics must be acknowledged and explained.

Haught tries to do just that by constructing a complicated system based on the idea of an autonomous creation and a humble Creator. What science detects as evolution, according to Haught, is really the world's ongoing process of self-creation. The world is "self-ordering" and "self-creative" as it moves "into an always free and open future."[33] This process appears random and unguided to evolutionists because God does not control his creation, for true love is not coercive but persuasive. God has more of a vision than a fixed plan as he entices the world with opportunity rather than manipulating the world with smothering love.

The Scriptures speak of God's rescuing a fallen world, but for Haught the human sin condition is not particularly threatening. The sin so deplored in Scripture is not so much our actual evil acts as the "intractable situation that has come to prevail" as a result of humanity's "indifference to its creative mission in the cosmos."[34] In Haught's system the biblical doctrine of original sin refers to how "each of us is born into a still unfinished, imperfect universe."[35] Therefore God need not be seen as the take-charge sort of God who rescues the world. Instead the Christian view is better represented by a vulnerable, defenseless Creator who respects the world. "God's unobtrusive and self-absenting mode of being," explains Haught, "invites the world to swell forth continually."[36] It is, paradoxically, the "hiddenness of God's power in a self-effacing persuasive love, that allows creation to come about and to unfold freely and indeterminately in evolution."[37] In fact, God's love is so "outrageously 'irrational'" that Haught wonders if intelligence might have first arisen in the evolution of humankind.[38] In Haught's system, God is in "humble retreat." He is unobtrusive, self-concealing, self-withdrawing, and under eternal restraint.

Haught has rescued God from evil, not by moving God upward toward transcendence but by moving him downward toward subservience. God's humility and selflessness are accompanied by a loss of control and responsibility. This is exactly what we should expect, argues Haught, because this is how true love works.

One result of Haught's system is that science need not concern itself with the actions of God, for his account "in no way interferes with purely scientific explanations of evolutionary events." The "God hypothesis,"[39] Haught reassures the scientist, need not be considered. This, of course, could also be said of the ideas of Dobzhansky, Gray, Van Till, and Miller. But this is old news—the "God hypothesis" was rejected centuries ago. From Burnet in the seventeenth century to Lyell in the nineteenth century, otherwise religious thinkers sought to describe the world without resorting to God's direct actions. This concept of God served as the foundation for evolution, and with the acceptance of evolution, that concept continues to be foundational in today's thought.

The various attempts to reconcile God and evolution examined above reveal the refinement of seventeenth- and eighteenth-century theologies. The pre-Darwinian metaphysic of a Creator with no direct influence on the world becomes in our day much the same thing—a Creator who is not manifest in the world. Whether the attempt is toward the transcendent deity of Gnosticism or the subdued deity of process theology, the effect is similar. Whether in one direction or another, God is distanced from the world and, more important, from its evil.

In this brief survey of theistic evolutionists, we have seen that Dobzhansky, Gray, Van Till, Miller, and Haught all accept and even rely on the Darwinian type of metaphysical arguments against the view that a divine hand is active in creating and sustaining the world. Warfield, who was willing to accommodate evolution but was not committed to the idea, did not align himself with the evolution theodicy. Gray and Van Till, on the other hand, may be examples of how difficult it is to maintain commitments to a God who is in control of the world and to Darwin's theory of evolution. It is not surprising that science historian John Hedley Brooke concluded that beliefs in evolution and in a sovereign God do not overlap: "It should not be difficult to see why intelligent people have often taken the view that Darwin's theory, properly understood, and Christian conceptions of an active providence are not merely incompatible but belong to two mutually exclusive worlds of thought."[40]

Darwin constructed his theory of evolution to explain the quandaries of the natural world. He believed that God could not be responsible for nature's carnage and inefficiency, so he proposed a purely naturalistic explanation. Evolution was a theodicy, and keeping this in mind helps explain the different responses to evolution, including those critics such as Hodge and the theistic evolutionists. This perspective also helps explain how those who accept evolution wholeheartedly can be content with evidence that establishes merely the plausibility of evolution. We saw this repeatedly in chapters 2 through 5.

Over and over the evidence presented fails to give a compelling argument for why evolution must have occurred; instead it supports the mere plausibility of evolution. The theory of evolution makes the high claim that the most complex things we know of—living organisms—arose from blind forces of nature. We are told that complexity and even consciousness just bubbled up out of an inorganic world. These are extraordinary claims and therefore they require extraordinary evidence. Instead we have a series of unsubstantiated speculations. These speculations are made compelling, however, by evolution's negative theology.

Arguments aimed at proving evolution are characterized by a constant switching from scientific speculation to metaphysical pronouncement. Darwin admitted that his theory could only to a certain extent explain homologies, but after giving his metaphysical interpretation of homologies he confidently stated that he would without hesitation adopt evolution even if there were no other facts or arguments. From a scientific perspective homologies could be used only to show how evolution might have happened, but from a metaphysical perspective they could make evolution compelling.

Who Is Your God?

The National Academy of Sciences' document *Science and Creationism* states that only science should be taught in science classes:

> No body of beliefs that has its origin in doctrinal material rather than scientific observation, interpretation, and experimentation should be admissible as science in any science course. Incorporating the teaching of such doctrines into a science curriculum compromises the objectives of public education.[41]

In saying this, the evolutionists become their own judge. The only possible conclusion is that evolution should not be taught in science classes, for Darwin's theory goes far beyond "scientific observation, interpretation, and experimentation." It includes religious presuppositions outside of science. Evolutionists argue that homologies and small-scale changes in species can only be explained by evolution, and that the fossil record makes evolution a fact. Evolutionists come to these conclusions because they believe in a certain type of God and creation—beliefs that are not open to scientific debate.

It is important to understand that evolution relies on religious premises, but it is even more important to understand that evolutionists do not acknowledge this reliance on metaphysical ideas. An unspoken, unscientific position underlies evolution, and until this is understood public debate will continue to be more confusing than enlightening. For a fruitful public debate, we need to understand evolution's foundation. We need to understand the metaphysical interpretations that are attached to the scientific observations. We need to understand these things because, ultimately, evolution is not about the scientific details. Ultimately, evolution is about God.

Notes

Chapter 1

1. Neal C. Gillespie, *Charles Darwin and the Problem of Creation* (Chicago: University of Chicago Press, 1979), 72, 77–79, 126–27.

2. Adrian Desmond and James Moore, *Darwin* (New York: W. W. Norton, 1991), 449.

3. Ernst Mayr, *Toward a New Philosophy of Biology* (Cambridge, Mass.: Harvard University Press, Belknap Press, 1988), 170.

4. Charles Darwin, *The Autobiography of Charles Darwin* (New York: Harcourt, Brace and Company, 1958), 85.

5. A. N. Wilson, *God's Funeral* (New York: W. W. Norton, 1999), 192.

6. See for example: Karen Armstrong, *A History of God* (London: Heinemann, 1992), 362, 335–36; and Stanley J. Grenz and Roger E. Olson, *Twentieth-Century Theology: God and the World in a Transitional Age* (Downers Grove, Ill.: InterVarsity Press, 1992), 18–23.

7. Quoted in Steven Jay Gould, "Nonmoral Nature," in *Hen's Teeth and Horse's Toes* (New York: W. W. Norton, 1983).

8. Gillespie, *Charles Darwin and the Problem of Creation*, 72, 77–79, 126–27.

9. Charles Darwin, *The Origin of the Species*, 6th ed. (1872; repr. London: Collier Macmillan, 1962), 467–72. While the author is aware that the original title of Darwin's work was *On the Origin of Species*, since the reprint being cited is entitled *The Origin of Species*, it will be referred to as such throughout the book.

10. "We can see why throughout nature the same general end is gained by an almost infinite diversity of means, for every peculiarity when once acquired is long inherited, and structures already modified in many different ways have to be adapted for the same general purpose. We can, in short, see why nature is prodigal in variety, though niggard in innovation. But why this should be a law of nature if each species has been independently created no man can explain" (ibid., 468–69).

11. For example, Darwin wrote: "How curious it is, to give a subordinate though striking instance, that the hind-feet of the kangaroo, which are so well fitted for bounding over the open plains,— those of the climbing, leaf eating koala, equally well fitted for grasping the branches of trees,—those of the ground-dwelling, insect or root-eating, bandicoots,—and those of some other Australian

marsupials,—should all be constructed on the same extraordinary type, namely with the bones of the second and third digits extremely slender and enveloped within the same skin, so that they appear like a single toe furnished with two claws. Notwithstanding this similarity of pattern, it is obvious that the hind feet of these several animals are used for as widely different purposes as it is possible to conceive. . . . Nothing can be more hopeless than to attempt to explain this similarity of pattern in members of the same class, by utility or by the doctrine of final causes" (ibid., 434–35).

12. Adam Sedgwick, *A Discourse on the Studies of the University* (London: John W. Parker, 1833), 51.

13. Ibid., 23.

14. Ibid., 19.

15. Ibid., 14.

16. Job 38:3–21; 39:13–17; Romans 1:19–20; 8:20–22; Psalm 19:1; Isaiah 45:7.

17. Charles Darwin, *Life and Letters of Charles Darwin*, ed. Francis Darwin (London: John Murray, 1887), 2:249.

18. Ibid.

19. Darwin, *Autobiography*, 90.

Chapter 2

1. For our purposes, the term *homology* will be used to refer to derived homologies as opposed to ancestral homologies, and I will ignore the difficulty in distinguishing monophyletic from paraphyletic groups.

2. "I should without hesitation adopt [evolution], even if it were unsupported by other facts or arguments" (Darwin, *Origin*, 457).

3. See, for example, Mark Ridley, *Evolution* (Boston: Blackwell Scientific, 1993), chap. 3; and Stephen Jay Gould, "Darwinism Defined: The Difference between Fact and Theory," *Discover*, January 1987.

4. Ridley, *Evolution*, 44.

5. Ibid., 50.

6. Ibid.

7. Quoted in Darwin, *Origin*, 440.

8. Ibid., 441.

9. Ibid.

10. Ibid., 443.

11. Ibid., 449.

12. Ernst Haeckel, *General Morphology of Organisms* (Berlin: Georg Reimer, 1866).

13. Stephen Jay Gould, "SETI and the Wisdom of Casey Stengel," in *The Flamingo's Smile* (New York: W. W. Norton, 1985), 413.

14. Sir Julian Huxley, *Evolution in Action* (New York: Signet Science, 1953), 16–19.

15. Tim M. Berra, *Evolution and the Myth of Creationism* (Stanford, Calif.: Stanford University Press, 1990), 22.

16. See, for example, Ridley, *Evolution*, 46–48.

17. Emile Zuckerkandl and Linus Pauling, "Molecules as Documents of Evolutionary History" *Journal of Theoretical Biology* 8 (1965): 357–66.

18. National Academy of Sciences, *Science and Creationism: A View from the National Academy of Sciences*, 2d ed. (Washington, D.C.: National Academy Press, 1999), 19.

19. David Penny, L. R. Foulds, and M. D. Hendy, "Testing the Theory of Evolution by Comparing Phylogenetic Trees Constructed from Five Different Protein Sequences," *Nature* 297 (1982): 197–200.

20. Ridley, *Evolution*, 50–52.

21. Alberts and Ayala, *Science and Creationism*, 20.

22. Michael Ruse, *Taking Darwin Seriously* (New York: Basil Blackwell, 1986), 4.

23. Christian de Duve, *Vital Dust* (New York: Basic Books, 1995), 1.

24. Kenneth R. Miller, review of *Darwin's Black Box: The Biochemical Challenge to Evolution* by Michael J. Behe, *Creation/Evolution* 16 (2): 36–40.

25. Niles Eldredge, *The Monkey Business* (New York: Washington Square, 1982), 41.

26. Ridley, *Evolution*, 454.

27. Ibid., 456 (emphasis added).

28. Darwin, *Origin*, 437 (emphasis added).

29. Berra, *Evolution and the Myth*, 64–66.

30. See, for example, Alan Fersht, *Structure and Mechanism in Protein Science* (New York: W. H. Freeman, 1999), 26–30; and Carl Branden and John Tooze, *Introduction to Protein Structure*, 2d ed. (New York: Garland, 1999), 208–19.

31. Berra, *Evolution and the Myth*, 66.

32. Ibid.

33. S. R. Scadding, "Do Vestigial Organs Provide Evidence for Evolution?" *Evolutionary Theory* 5 (1981): 173–76.

34. Edward O. Dodson and Peter Dodson, *Evolution: Process and Product* (New York: D. Van Nostrand, 1976), 51.

35. Berra, *Evolution and the Myth*, 22.

36. Ridley, *Evolution*, 50.

37. Berra, *Evolution and the Myth*, 22.

38. A. S. Romer and T. S. Parsons, *The Vertebrate Body*, 6th ed. (Philadelphia: Saunders College Publishing, 1986), 407; quoted in W. R. Bird, *The Origin of Species Revisited* (Nashville: Regency, 1991), 1:274.

39. Hubert P. Yockey, *Information Theory and Molecular Biology* (Cambridge: Cambridge University Press, 1992), 172–77.

40. Florence Raulin-Cerceau et al., "From Panspermia to Bioastronomy: The Evolution of the Hypothesis of Universal Life," *Origins of Life and Evolution of the Biosphere* 28 (1998): 597–612.

41. See, for example, Jacques Ninio, *Molecular Approaches to Evolution* (Princeton, N.J.: Princeton University Press, 1983), 79–81; Hyman Hartman, "Speculations on the Evolution of the Genetic Code," *Origins of Life and Evolution of the Biosphere* 25 (1995): 265.

42. See, for example, Leslie E. Orgel, "The Origin of Life—How Long Did It Take?" *Origins of Life and Evolution of the Biosphere* 28 (1998): 91–96; Mitchell K. Hobish, "Studies on Order in Prebiological Systems at the Laboratory of Chemical Evolution," *Origins of Life and Evolution of the Biosphere* 28 (1998): 124; Ninio, *Molecular Approaches*, 89.

43. The vertebrate mitochondria were found to use a slightly different genetic code; see B. G. Barrell et al., "A Different Genetic Code in Human Mitochondria," *Nature* 282 (1979): 189–94.

44. Syozo Osawa, *Evolution of the Genetic Code* (Oxford: Oxford University Press, 1995), 71–72.

45. Carl R. Woese, "Archaebacteria," *Scientific American* 244, no. 6 (1981): 98–122.

46. Carl R. Woese, "The Universal Ancestor," *Proceedings of the National Academy of Science* 95, no. 12 (1998): 6854–59.

47. Penny, Foulds, and Hendy, "Testing the Theory of Evolution."

48. Alan F. Chalmers, *What Is This Thing Called Science?* 2d ed. (Indianapolis: Hackett, 1982), 60–75.

49. Christian Schwabe writes, "Evolutionary trees constructed from different proteins suggest the existence of different genealogies instead of a unique one," in "On the Validity of Molecular Evolution," *Trends in Biochemical Sciences* 11 (July 1986): 280–82. See also Donald L. J. Quicke, *Principles and Techniques of Contemporary Taxonomy* (London: Blackie Academic and Professional, 1993), 34–35.

50. See, for example, Peter J. Andrews, "Aspects of Hominoid Phylogeny," in *Molecules and Morphology in Evolution*, ed. Colin Patterson (Cambridge: Cambridge University Press, 1987), 28.

51. Thomas H. Jukes and Richard Holmquist, "Evolutionary Clock: Nonconstancy of Rate in Different Species," *Science* 177 (1972): 530–32.

52. R. P. Ambler and Margaret Wynn, "The Amino Acid Sequences of Cytochromes c–551 from Three Species of *Pseudomonas*," *Biochemical Journal* 131 (1973): 485–98; and R. P. Ambler et al.,

"Cytochrome c Sequence Variation among the Recognised Species of Purplee Nonsulphur Photosynthetic Bacteria," *Nature* 278 (1979): 659–60.

53. Christian Schwabe and Gregory W. Warr, "A Polyphyletic View of Evolution: The Genetic Potential Hypothesis," *Perspectives in Biology and Medicine* 27 (1984): 465–78; and Schwabe, "On the Validity of Molecular Evolution."

54. Schwabe and Warr, "Polyphyletic View of Evolution."

55. Gabriel A. Dover, "DNA Turnover and the Molecular Clock," *Journal of Molecular Evolution* 26 (1987): 47–58.

56. Schwabe, "On the Validity of Molecular Evolution."

57. Peter E. M. Gibbs et al., "The Molecular Clock Runs at Different Rates among Closely Related Members of a Gene Family," *Journal of Molecular Evolution* 46 (1998): 552–561.

58. Francisco J. Ayala, "Molecular Clock Mirages," *BioEssays* 21 (1999): 71–75.

59. Michael S. Y. Lee, "Molecular Clock Calibrations and Metazoan Divergence Dates," *Journal of Molecular Evolution* 49 (1999): 385–91.

60. Darwin, *Origin*, 457.

61. Ridley, *Evolution*, 49.

62. Berra, *Evolution and the Myth*, 19.

63. Ridley, *Evolution*, 52–54.

64. Darwin, *Origin*, 468.

65. Douglas J. Futuyma, *Science on Trial* (New York: Pantheon, 1983), 55.

66. Darwin, *Origin*, 434–35.

67. Ibid., 437.

68. Stephen Jay Gould, "The Panda's Thumb," in *The Panda's Thumb* (New York: W. W. Norton, 1980), 20.

69. Quoted in Michael J. Behe, *Darwin's Black Box: The Biochemical Challenge to Evolution* (New York: Free Press, 1996), 225–26.

70. Gould, "Panda's Thumb," 20–21.

71. Ridley, *Evolution*, 50.

72. Gould, "Darwinism Defined."

73. Berra, *Evolution and the Myth*, 22.

74. Futuyma, *Science on Trial*, 46, 48, 62, 199.

Chapter 3

1. See, for example, Z. Lu et al., "Evolution of an *Escherichia coli* Protein with Increased Resistance to Oxidative Stress," *Journal of Biological Chemistry* 273, no. 14 (April 3, 1998): 8308–16; and A. J. Hacking et al., "Evolution of Propanediol Utilization in *Escherichia coli*: Mutant with Improved Substrate-Scavenging Power," *Journal of Bacteriology* 136, no. 2 (November 1978): 522–30.

2. Ernst Mayr, *Toward a New Philosophy of Biology* (Cambridge, Mass.: Harvard University Press, Belknap Press, 1988), 208.

3. Darwin, *Origin*, 150.

4. Ibid., 150–51.

5. Ibid., 93.

6. Ibid.

7. Mayr, *Toward a New Philosophy*, 425.

8. Sir Julian Huxley, *Evolution in Action* (New York: Signet Science, 1953), 40.

9. Mark Ridley, *Evolution* (Boston: Blackwell Scientific, 1993), 43–44.

10. Niles Eldredge, *The Monkey Business* (New York: Washington Square, 1982), 51.

11. Ridley, *Evolution*, 522.

12. Mayr, *Toward a New Philosophy*, 192.

13. Isaac Asimov, *Asimov's New Guide to Science* (New York: Basic Books, 1984).

14. Steve Jones, *Darwin's Ghost* (New York: Random House, 2000), 15.

15. Jonathan Weiner, "Kansas Anti-evolution Vote Denies Students a Full Spiritual Journey," *Philadelphia Enquirer*, August 15, 1999.

16. John Hedley Brooke, *Science and Religion: Some Historical Perspectives* (Cambridge: Cambridge University Press, 1991), 197, 231–32.

17. "In particular, it was Darwin's realization of the invalidity of three prominent doctrines among the numerous beliefs of creationism that was of crucial importance for Darwin's change of mind: (1) that of an unchanging world of short duration, (2) that of the constancy of sharp delimitation of created species, and (3) that of a perfect world explicable only by the postulate of an omnipotent and beneficent Creator. . . . Darwin's conclusions were reinforced by his conviction of the invalidity of several other tenets of creationism, such as the special creation of man, but the three stated dogmas were clearly of primary importance" (Mayr, *Toward a New Philosophy*, 170).

18. Ernst Mayr, *The Growth of Biological Thought* (Cambridge, Mass.: Harvard University Press, Belknap Press, 1982), 403.

19. Darwin, *Origin*, 66.

20. Ibid., 66–67.

21. Quoted in Jonathan Weiner, *The Beak of the Finch* (New York: Vintage, 1995), 297.

22. David J. Merrell, *Evolution and Genetics* (New York: Holt, Rinehart and Winston, 1962), 162.

23. Weiner , "Kansas Anti-evolution Vote."

24. Ridley, *Evolution*, 37–59.

25. Berra, *Evolution and the Myth*, 55.

26. Mayr, *Toward a New Philosophy*, 192.

Chapter 4

1. J. G. M. Thewissen et al., "Fossil Evidence for the Origin of Aquatic Locomotion in Archaeocete Whales," *Science* 263 (1994): 210–12.

2. Stephen J. Gould, "Hooking Leviathan by Its Past," *Natural History*, May 1994.

3. Mark Ridley, *Evolution* (Boston: Blackwell Scientific, 1993), 53–54.

4. Ibid., 514.

5. Ibid., 516.

6. National Academy of Sciences, *Science and Creationism*, 14.

7. Charles Darwin, *The Life and Letters of Charles Darwin*, 3:248; quoted in Michael Denton, *Evolution: A Theory in Crisis* (Bethesda, Md.: Adler and Adler, 1985), 163.

8. T. S. Kemp, *Fossils and Evolution*, (Oxford: Oxford University Press, 1999), 16.

9. George G. Simpson, *The Meaning of Evolution*, rev. ed. (New Haven: Yale University Press, 1967), 113–14.

10. Stephen Jay Gould, "Is the Cambrian Explosion a Sigmoid Fraud?" in *Ever Since Darwin: Reflections in Natural History* (New York: W. W. Norton, 1973).

11. Steve Jones, *Darwin's Ghost* (New York: Random House, 2000), 206.

12. Regarding the changes observed in the guppies, David Reznick writes: "The rate and patterns of change attainable through natural selection are sufficient to account for the patterns observed in the fossil record" (see David Reznick et al., "Evaluation of the Rate of Evolution in Natural Populations of Guppies [*Poecilia reticulata*]," *Science* 275 [1997]: 1934–36). Quoted in Kenneth R. Miller, *Finding Darwin's God* (New York: Cliff Street Books, 1999), 111.

13. Eldredge, *Monkey Business*, 47.

14. Jones, *Darwin's Ghost*, 207.

15. Lisa J. Shawver, "Trilobite Eyes: An Impressive Feat of Early Evolution," *Science News* 105 (1974): 72.

16. Riccardo Levi-Setti, *Trilobites*, 2d ed. (Chicago: University of Chicago Press, 1993), 29.

17. Gordon Rattray Taylor, *The Great Evolution Mystery* (New York: Harper & Row, 1983), 14.

18. Robert Wesson, *Beyond Natural Selection* (Cambridge, Mass.: MIT Press, 1991), 73.

19. Michael J. Behe, *Darwin's Black Box: The Biochemical Challenge to Evolution* (New York: Free Press, 1996), 39.

20. Richard Dawkins, *The Blind Watchmaker* (New York: W. W. Norton, 1986), 87.

21. For example, Christopher Gregory Weber, "The Bombardier Beetle Myth Exploded," *Creation/Evolution* 3 (1): 1–5; and Christopher Gregory Weber, "Response to Dr. Kofahl," *Creation/Evolution* 5 (2): 15–17.

22. Darwin, *Origin*, 178.

23. Ibid.

24. Ibid., 201–2.

25. Ibid., 182 (emphasis added).

26. For example, the Krebs cycle is a complicated and apparently optimal metabolic pathway that has been a problem for evolutionists to explain. One recent paper claims to demonstrate the "opportunistic evolution of the Krebs cycle," but what passes for a demonstration is really a series of speculations about what might have happened, with no actual details of the particulars. See Enrique Melendez-Hevia et al., "The Puzzle of the Krebs Citric Acid Cycle," *Journal of Molecular Evolution* 43 (1996): 293–303.

27. Michael J. Novacek, "Whales Leave the Beach," *Nature* 368 (1994): 807.

28. See Stephen Jay Gould, "Life's Little Joke," in *Bully for Brontosaurus* (New York: W. W. Norton, 1991); and Douglas J. Futuyma, *Evolutionary Biology* (Sunderland, Mass.: Sinauer Associates, 1986).

29. Eldredge, *Monkey Business*, 75.

30. Laurie R. Godfrey, "Creationism and Gaps in the Fossil Record," in *Scientists Confront Creationism* (New York: W. W. Norton, 1983) (emphasis in original).

31. Futuyma, *Science on Trial*; quoted in Phillip Johnson, *Darwin on Trial* (Downers Grove, Ill.: InterVarsity Press, 1991), 76.

32. T. S. Kemp, *Mammal-like Reptiles and the Origin of Mammals*, (London: Academic Press, 1982), 326–27; and Robert L. Carroll, *Vertebrate Paleontology and Evolution*, (New York: W. H. Freeman and Company, 1988), 397.

33. Alfred S. Romer, *Vertebrate Paleontology*, 3d ed., (Chicago: University of Chicago Press, 1966), 184–85; George G. Simpson in: Carlotta Kerwin et. al., *Life Before Man* (New York: Time-Life Books, 1972), 42; Carroll, *Vertebrate Paleontology and Evolution*, 397–98.

34. Michael J. Benton, *Vertebrate Paleontology*, 2d ed. (London: Chapman and Hall, 1997), 291; Romer, *Vertebrate Paleontology*, 184; Carroll, *Vertebrate Paleontology and Evolution*, 377.

35. Niles Eldredge and Stephen Jay Gould, "Punctuated Equilibrium Prevails," *Nature* 332 (1988): 211–12.

36. Niles Eldredge, "An Extravagance of Species," *Natural History* (American Museum of Natural History) 89, no. 7 (1980): 50.

37. Quoted in Stephen Jay Gould, "The Episodic Nature of Evolutionary Change," in *The Panda's Thumb* (New York: W. W. Norton, 1980), 179.

38. Gillespie, *Charles Darwin and the Problem of Creation*, 119.

39. Quoted in Gould, "Episodic Nature," 181.

40. Ibid., 184.

41. Richard A. Kerr, "Did Darwin Get It All Right?" *Science* 267 (1995): 1421–22.

42. Berra, *Evolution and the Myth*, 31.

43. Ibid., 50.

44. George G. Simpson, *Horses* (Oxford: Oxford University Press, 1951); quoted in Richard Milton, *Shattering the Myths of Darwinism* (Rochester, Vt.: Park Street Press, 1992), 102.

45. Miller, *Finding Darwin's God*, 43.

46. Berra, *Evolution and the Myth*, 142.

47. Martin Gardner, "Denying Darwin," *Commentary*, September 1996, 16.

48. Quoted in Antony Flew, *Darwinian Evolution*, 2d ed. (New Brunswick, N.J.: Transaction Publishers, 1997), 18.

49. Gavin de Beer, *Atlas of Evolution* (London: Nelson, 1964), 48.

50. Miller, *Finding Darwin's God*, 102.

51. Futuyma, *Science on Trial*, 80 (emphasis added).

52. Berra, *Evolution and the Myth*, 39 (emphasis in original).

53. Ridley, *Evolution*, 56.

54. Gould, "Darwinism Defined".

55. Miller, *Finding Darwin's God*, 41.

56. Futuyma, *Science on Trial*, 127.

57. Miller, *Finding Darwin's God*, 97.

58. Ibid., 100–3.

59. Jones, *Darwin's Ghost*, 130–31.

60. Romans 8:20.

Chapter 5

1. Joseph Le Conte, *Evolution: Its Nature, Its Evidences, and Its Relation to Religious Thought*, 2d ed. (New York: D. Appleton, 1891), 54.

2. Ibid., 55.

3. Ibid., 56.

4. Ibid., 55.

5. Ibid.

6. Almost a century later, in 1982, Ernst Mayr wrote, "The concept of the infinity of the universe became increasingly accepted, and this process has continued right up to modern astronomy" (Ernst Mayr, *The Growth of Biological Thought* [Cambridge, Mass.: Harvard University Press, Belknap Press, 1982], 313).

7. Le Conte, *Evolution*, 54–55.

8. Ibid., 65.

9. Ibid., 65–66.

10. Ibid., 66.

11. Mark Ridley, *Evolution* (Boston: Blackwell Scientific, 1993), 57, 323.

12. Le Conte, *Evolution*, 67.

13. Ibid.

14. Ibid., 80.

15. Ibid., 107.

16. Ibid., 128–29.

17. Ibid., 145 (emphasis in original).

18. Ibid., 180.

19. Ibid., 162.

20. Ibid., 194.

21. H. H. Lane, *Evolution and Christian Faith* (Princeton, N.J.: Princeton University Press, 1923), 31.

22. Job 39:14–17.

23. Lane, *Evolution and Christian Faith*, 32.

24. Ibid. (emphasis added).

25. Ibid., 33.

26. Ibid., 38.

27. Ibid., 41.

28. Ibid., 45.

29. Ibid., 47 (emphasis in original).

30. For example, after discussing the groupings of species, Darwin concludes that "these are strange relations on the view that each species was independently created" and that it was "utterly inexplicable on the theory of creation" (Darwin, *Origin*, 468).

31. Berra, *Evolution and the Myth*, 19–20.

32. Edward O. Dodson and Peter Dodson, *Evolution: Process and Product* (New York: D. Van Nostrand, 1976), 65.

33. Arthur W. Lindsey, *Principles of Organic Evolution* (St. Louis, Mo.: C. V. Mosby Company, 1952), 123.

34. Ibid., 112.

35. Ibid., 114.

36. Ibid., 115 (emphasis added).

37. Ibid., 116.

38. Ibid., 118.

39. Ibid., 119–20 (emphasis added).

40. Ibid., 120.

41. Ibid., 120, 133–34.

42. De Beer, *Atlas of Evolution*, 5.

43. Ibid.

44. Ibid., 34–35.

45. Ibid., 35.

46. Ibid., 34.

47. Ibid., 35.

48. Ibid., 38.

49. Ibid.

50. Ibid.

51. Ibid.

52. Ibid.

53. Ibid., 39.

54. Ibid., 40.

55. Ibid.

56. Ibid., 43.

57. Ibid., 44.

58. Ridley, *Evolution*, 377–80.

59. George S. Carter, *A Hundred Years of Evolution* (London: Sidgwick and Jackson, 1957), 15.

60. Niles Eldredge, *The Triumph of Evolution and the Failure of Creationism* (New York: W. H. Freeman, 2000), 27.

61. Ibid., 30–31.

62. Ibid., 146.

63. See, for example, Leon Croizat et al., "Centers of Origin and Related Concepts," *Systematic Zoology* 23 (1974): 271–73; Gareth Nelson, "From Candolle to Croizat: Comments on the History of Biogeography," *Journal of the History of Biology* 11 (1978): 288–96.

64. For example, Mark Ridley has an entire chapter on biogeography in his evolution textbook but does not use biogeography as one of his evidences for evolution (Ridley, *Evolution*).

65. C. Patterson, "Aims and Methods in Biogeography," in Systematics Association Special Volume No. 23, *Evolution, Time and Space: The Emergence of the Biosphere*, ed. R. W. Sims et al. (London and New York: Academic, 1983), 9.

66. Alec L. Panchen, *Classification, Evolution, and the Nature of Biology* (Cambridge: Cambridge University Press, 1992), 103–4.

67. De Beer, *Atlas of Evolution*, 48.

68. Ibid.

69. Verne Grant, *The Evolutionary Process*, 2d ed. (New York: Columbia University Press, 1991), 13.

70. Ibid.

71. Ibid., 14.

72. Ibid., 12.

73. Moody writes, "Clearly, the burden of proof lies with the affirmative in the matter of proving the usefulness of vestiges for which no functions have ever been discovered" (see Paul A. Moody, *Introduction to Evolution*, 3d ed. [New York: Harper & Row, 1970], 42.

74. Michael Ruse, *Darwinism Defended* (Reading, Mass.: Addison-Wesley, 1982), 40.

75. Dodson and Dodson, *Evolution*, 29.

Chapter 6

1. Quoted in Charles C. Gillispie, *Genesis and Geology* (Cambridge, Mass.: Harvard University Press, 1951), xxviii.

2. Ibid., 6–7.

3. Thomas Huxley's essay "Evolution in Biology" is quoted in John C. Greene, *Science, Ideology, and World View* (Berkeley: University of California Press, 1981), 141.

4. Quoted in Steven Jay Gould, *Ever Since Darwin: Reflections in Natural History* (New York: W. W. Norton, 1973), 141–46.

5. Peter J. Bowler, *Evolution: The History of an Idea* (Berkeley: University of California Press, 1984), 28.

6. Psalm 62:12.

7. Quoted in Colin Brown, *Philosophy and the Christian Faith* (Downers Grove, Ill.: InterVarsity Press, 1968), 57.

8. Galatians 6:7.

9. Francis J. Beckwith, *David Hume's Argument against Miracles: A Critical Analysis* (Lanham, Md.: University Press of America, 1989), 28–32.

10. Quoted in A. N. Wilson, *God's Funeral* (New York: W. W. Norton, 1999), 185.

11. Beckwith, *David Hume's Argument*, 32.

12. Rudolf Bultmann, "Is Exegesis without Presupposition Possible?" *Existence and Faith* (New York: Meridian Books, 1960), 291; quoted in John Wenham, *Redating Matthew, Mark, and Luke* (Downers Grove, Ill.: InterVarsity Press, 1992), 248.

13. Charles Darwin, *The Autobiography of Charles Darwin* (New York: Harcourt, Brace and Company, 1958), 87.

14. "Such to perfection, one first matter all, / Indued with various forms, various degrees, . . . Differing but in degree, of kind the same, . . . Improved by tract of time, and winged ascend" (John Milton, *Paradise Lost* [repr., New York: W. W. Norton, 1975], 5:469–503).

15. Isaiah 45:7.

16. Of Milton's doctrine of creation, his great biographer David Masson wrote: "God has voluntarily loosened his hold on such portions of this primeval matter as he has endowed with free will, so that they may originate independent actions not morally referable to God himself." Quoted in Augustus H. Strong, *The Great Poets and Their Theology* (Philadelphia: American Baptist Publication Society, 1897), 263.

17. Darwin, *Origin*, 483–84.

18. Specifically, this passage refers to the ideas of Ralph Cudworth and Henry More.

19. See Richard S. Westfall, *Science and Religion in Seventeenth-Century England* (Ann Arbor: University of Michigan Press, 1958), 94.

20. Philip J. Lee, *Against the Protestant Gnostics* (Oxford: Oxford University Press, 1987), 16.

21. John Hedley Brooke, *Science and Religion* (Cambridge: Cambridge University Press, 1991), 144ff.

22. For a contemporary survey of the problem of evil, including this problem of too much evil—the inductive argument from evil—see Bruce R. Reichenbach, *Evil and a Good God* (New York: Fordham University Press, 1982).

23. Quoted in John F. Cornell, "God's Magnificent Law: The Bad Influence of Theistic Metaphysics on Darwin's Estimation of Natural Selection," *Journal of the History of Biology* 20, no. 3 (1987): 390.

24. Nehemiah Grew, *Cosmologia Sacra*, 1701; quoted in Westfall, *Science and Religion in Seventeenth-Century England*, 60.

25. David Hume, *Dialogues concerning Natural Religion* (Oxford: Clarendon Press, 1779).

Chapter 7

1. There were, of course, exceptions to this generalization. For example, there were scientists such as the Swiss geologist Jean Andre Deluc who moved to London in 1773, and the French anatomist

Baron Cuvier, and popularizers such as the Duke of Argyll and Hugh Miller who held to more God-centered religious views. There was even a revival of Calvinism at Oxford in the 1820s.

2. A. N. Wilson, *God's Funeral* (New York: W. W. Norton, 1999), 46.

3. Erasmus Darwin, *Zoonomia; or The Laws of Organic Life*, vol. 1 (London: J. Johnson, 1794), 509; quoted in George B. Dysan, "Darwin in Kansas," *Science* 285 (1999): 1355.

4. Michael Horton, "The New Gnosticism," *Modern Reformation*, July/August 1995, 4–12.

5. John 1:14.

6. Wilson, *God's Funeral*, 129–30.

7. Philip J. Lee, *Against the Protestant Gnostics* (Oxford: Oxford University Press, 1987), 17.

8. William Paley, *The Principles of Moral and Political Philosophy*, 20th ed., vol. I (London: J. Faulder, 1814), 71.

9. Adrian Desmond and James Moore, *Darwin* (New York: W. W. Norton, 1991), 78.

10. Quoted in Steve Jones, *Darwin's Ghost* (New York: Random House, 2000), 128.

11. Charles Lyell, *Principles of Geology* (London: Penguin, 1997), 437.

12. Job 39:14–17.

13. Michael Ruse, *Darwinism Defended* (Reading, Mass.: Addison-Wesley, 1982), 27.

14. John 14:11.

15. Gillispie, *Genesis and Geology*, 195–96.

16. Ibid., 190.

17. Ibid., 209.

18. Theologian and geologist William Buckland (1784–1856); see Steven Jay Gould, "Non-moral Nature," in *Hen's Teeth and Horse's Toes* (New York: W. W. Norton, 1983), 32–43.

19. William Buckland, quoted in Gillispie, *Genesis and Geology*, 201.

20. Entomologist William Kirby (1759–1850); see Gould, "Nonmoral Nature."

21. Romans 8:22.

22. Quoted in Gillispie, *Genesis and Geology*, 195.

23. Quoted in Gillespie, *Charles Darwin*, 26.

24. William Coleman, *Biology in the Nineteenth Century* (New York: John Wiley and Sons, 1971), 62.

25. James Hutton, *Theory of the Earth* (Edinburgh, 1795); quoted in Gillispie, *Genesis and Geology*, 46.

26. Gillespie, *Charles Darwin*, 43.

27. Lyell, *Principles of Geology*, 80.

28. Steven Jay Gould states, "In fact, the catastrophists were much more empirically minded than Lyell. The geological record does seem to record catastrophes," in "Uniformity and Catastrophe," in *Ever Since Darwin: Reflections in Natural History* (New York: W. W. Norton, 1973), 150.

29. Quoted in Gillispie, *Genesis and Geology*, 126.

30. Quoted in ibid., 135 (emphasis added).

31. John C. Greene, *Science, Ideology, and World View* (Berkeley: University of California Press, 1981), 62–64.

32. Gillispie, *Genesis and Geology*, 190.

33. Greene, *Science, Ideology, and World View*, 72–73.

34. Quoted in ibid., 35–36.

35. Ibid., 63–66.

36. Quoted in Gillispie, *Genesis and Geology*, 153.

37. Quoted in Gillispie, *Charles Darwin*, 30–31.

38. Robert M. Young, *Darwin's Metaphor: Nature's Place in Victorian Culture* (Cambridge and New York: Cambridge University Press, 1985).

39. Sir Charles Sherrington, *Man on His Nature*, 2d ed. (Garden City, N.Y.: Doubleday/Anchor, 1955), 264–65.

40. George Shaw and Frederick P. Nodder, *The Naturalist's Miscellany: or, Coloured Figures of Natural Objects; Drawn and Described Immediately from Nature*, vol. X (London: Nodder & Co., 1799).

41. Alfred, Lord Tennyson, *In Memoriam A. H. H.*, in *The Oxford Anthology of English Literature*, vol. 2, 1800 to the Present, ed. Frank Kermode et al. (New York: Oxford University Press, 1973), 1239-40 (emphasis added).

42. Gillespie, *Charles Darwin*, 72, 77–79, 126–27.

43. Quoted in Desmond and Moore, *Darwin*, 449.

44. Quoted in Gould, "Nonmoral Nature."

45. Quoted in Greene, *Science, Ideology, and World View*, 138.

46. Ruse, *Darwinism Defended*, 3.

Chapter 8

1. Edgar Sheffield Brightman, quoted in Gordon H. Clark, *A Christian View of Men and Things* (Jefferson, Md.: Trinity Foundation, 1952), 275. Brightman saw this view as representing ancient thinkers such as Plato and modern thinkers such as David Hume, John Stuart Mill, William James, F. C. S. Schiller, H. G. Wells, Henri Bergson, William Montague, and others.

2. Gillispie, *Genesis and Geology*, 76.

3. Gillespie, *Charles Darwin*, 59.

4. Quoted in Gillispie, *Genesis and Geology*, 155.

5. Quoted in Jack Morrel and Arnold Thackray, *Gentlemen of Science* (Oxford: Clarendon, 1981), 236.

6. Baden Powell, *The Connexion of Natural and Divine Truth* (London, 1838), quoted in Robert M. Young, *Darwin's Metaphor: Nature's Place in Victorian Culture*, (Cambridge and New York: Cambridge University Press, 1985).

7. Quoted in Stephen Jay Gould, *Rocks of Ages* (New York: Ballantine, 1999), 203.

8. "I form the light and create darkness, I bring prosperity and create disaster," Isaiah 45:7; "I know every bird in the mountains, and the creatures of the field are mine," Psalm 50:11; "Look at the birds of the air; they do not sow or reap or store away in barns, and yet your heavenly Father feeds them," Matthew 6:26.

9. Darwin, *Origin*, 181.

10. Gould, *Rocks of Ages*, 4.

11. Eldredge, *Monkey Business*, 10.

12. National Academy of Sciences, *Science and Creationism*, ix.

13. From Joseph Le Conte's *Evolution: Its Nature, Its Evidences, and Its Relation to Religious Thought* to the National Academy of Sciences' *Science and Creationism: A View from the National Academy of Sciences*, evolutionists routinely appeal to mechanistic cosmogonies to argue that evolution is not limited to just life on earth but that the entire universe is the product of an evolutionary process.

14. Hans Jonas, quoted in Philip J. Lee, *Against the Protestant Gnostics* (Oxford: Oxford University Press, 1987), 16.

15. Harold Bloom, *The American Religion* (New York: Simon and Schuster, 1992), 22.

16. Gillespie, *Charles Darwin*, 151.

17. Quoted in ibid., 32–33 (emphasis added).

18. Darwin, *Origin*, 435.

19. Quoted in John C. Greene, *Science, Ideology, and World View* (Berkeley: University of California Press, 1981), 52.

20. Gillespie, *Charles Darwin*, 34.

21. Eldredge, *Monkey Business*, 39.

22. Paul A. Moody, *Introduction to Evolution*, 3d ed. (New York: Harper & Row, 1970), 26.

23. Tim M. Berra, *Evolution and the Myth of Creationism* (Stanford, Calif.: Stanford University Press, 1990), 66.

24. Ibid., 142.

25. Alan Donagan, "Can Anybody in a Post-Christian Culture Rationally Believe the Nicene Creed?" in *Christian Philosophy*, ed. Thomas P. Flint (Notre Dame, Ind.: University of Notre Dame Press, 1990), 110.

26. Maitland A. Edey and Donald C. Johanson, *Blueprints: Solving the Mystery of Evolution* (Boston: Little, Brown, 1989), 291.

27. Quoted in Gregg Easterbrook, "Science and God: A Warming Trend?" *Science* 277 (1997): 892 (emphasis added).

28. John Hedley Brooke, *Science and Religion* (Cambridge, Mass.: Cambridge University Press, 1991), 281.

29. See, for example, Frans de Waal, *Good Natured* (Cambridge, Mass.: Harvard University Press, 1991); and Robert Wright, *The Moral Animal* (New York: Pantheon, 1994).

30. Edward O. Wilson, *Consilience* (New York: Alfred A. Knopf, 1998), 262.

31. Ibid., 265.

32. David Hull, "The God of the Galápagos," *Nature* 352 (1991): 485–86.

33. Norman L. Geisler and Ronald M. Brooks, *Come Let Us Reason* (Grand Rapids: Baker, 1990), 95–96.

34. Ernst Mayr, *Toward a New Philosophy of Biology* (Cambridge, Mass.: Harvard University Press, Belknap Press, 1988), 192.

35. Quoted in Stephen Jay Gould, "Darwinism Defined: The Difference between Fact and Theory," *Discover*, January 1987.

36. Alfred North Whitehead, *Science and the Modern World* (New York: Macmillan, 1925), 49.

Chapter 9

1. Charles Hodge, *What Is Darwinism? And Other Writings on Science and Religion*, ed. Mark Noll and David Livingstone (Grand Rapids: Baker, 1994), 34–42.

2. Ibid.

3. Darwin, *Origin*, 182.

4. Hodge, *What Is Darwinism?* 95.

5. Theodosius Dobzhansky, *Mankind Evolving* (New Haven: Yale University Press, 1962), 6.

6. Theodosius Dobzhansky, *Evolution, Genetics, and Man* (New York: John Wiley and Sons, 1955), 374.

7. Quoted in Dobzhansky, *Mankind Evolving*, 347.

8. David N. Livingstone, *Darwin's Forgotten Defenders* (Grand Rapids: Eerdmans, 1987), 115–18.

9. For example, Dobzhansky argued for evolution by using metaphysical interpretations for a range of homologies in *Evolution, Genetics, and Man*, 227–28, 238, 244.

10. Livingstone, *Darwin's Forgotten Defenders*, 146.

11. See, for example, Dobzhansky, *Mankind Evolving*, 2; and Francisco J. Ayala and Walter M. Fitch, "Genetics and the Origin of Species: An Introduction," *Proceedings of the National Academy of Sciences* 94, no. 15 (1997): 7691–97.

12. Darwin, *Origin*, 468.

13. Terry Gray, "The Mistrial of Evolution: A Review of Phillip E. Johnson's *Darwin on Trial*," http://mcgraytx.calvin.edu/gray/evolution_trial/dotreview.html

14. Mark Ridley, *Evolution* (Boston: Blackwell Scientific, 1993), 52–54.

15. Gray, "Mistrial of Evolution."

16. Ibid.

17. Howard Van Till, "The Fully Gifted Creation," in *Three Views on Creation and Evolution*, ed. J. P. Moreland and John M. Reynolds (Grand Rapids: Zondervan, 1999), 167.

18. Howard Van Till, "God and Evolution: An Exchange," *First Things*, no. 34 (June/July 1993): 32–41.

19. Van Till, "Fully Gifted Creation," 187.

20. Ibid., 188.

21. Ibid., 246–47.

22. Kenneth R. Miller, *Finding Darwin's God* (New York: Cliff Street Books, 1999), 236.

23. Ibid., 237.

24. Ibid., 238–39.

25. Ibid., 239.

26. "Before I formed you in the womb I knew you, before you were born I set you apart," Jeremiah 1:5; "He chose us in him before the creation of the world to be holy and blameless in his sight," Ephesians 1:4.

27. Isaish 46:10–11.

28. Miller quotes process theologian Ian Barbour and professor John Polkinghorne, who holds to some parts of process theology. Miller makes use of ideas in process theology such as quantum indeterminacy and God's dependence on the events within creation. See Miller, *Finding Darwin's God*, chap. 8.

29. Ibid., 102.

30. Ibid., 238.

31. Ibid.

32. Ibid., 253.

33. John F. Haught, *God after Darwin* (Boulder, Co.: Westview, 2000), 53–54.

34. Ibid., 139.

35. Ibid., 138.

36. Ibid., 54.

37. Ibid., 97.

38. Ibid., 113.

39. Ibid., 53, 55.

40. John Hedley Brooke, *Science and Religion: Some Historical Perspectives* (Cambridge: Cambridge University Press, 1991), 305.

41. National Academy of Sciences, *Science and Creationism*, 25.

Index